Free Space Laser Communication with Ambient Light Compensation

Saleh Faruque

Free Space Laser Communication with Ambient Light Compensation

 Springer

Saleh Faruque
University of North Dakota
Grand Forks, ND, USA

ISBN 978-3-030-57486-4   ISBN 978-3-030-57484-0   (eBook)
https://doi.org/10.1007/978-3-030-57484-0

This Springer imprint is published by the registered company Springer Nature Switzerland AG
The registered company address is: Gewerbestrasse 11, 6330 Cham, Switzerland

Contents

Chapter 1
Introduction to Channel Coding

Topics
- Introduction to Channel Coding
- Types of Channel Coding
- Design Considerations
- Conclusions

1.1 Introduction to Channel Coding

Channel coding, also known as forward error control coding (FECC), is a process of detecting and correcting bit errors in digital communication systems. Channel coding is performed both at the transmitter and at the receiver [1–4]. Figure 1.1 shows the conceptual block diagram of a modern wireless communication system, where the channel coding block is shown in the inset of the dotted block. At the transmit side, channel coding is referred to as encoder, where redundant bits (parity bits) are added with the raw data before modulation. At the receive side, channel coding is referred to as the decoder. This enables the receiver to detect and correct errors, if they occur during transmission due to noise, interference, and fading. Since error control coding adds extra bits to detect and correct errors, transmission of coded information requires more bandwidth.

As the size and speed of digital data networks continue to expand, bandwidth efficiency becomes increasingly important. This is especially true for broadband communication, where the digital signal processing is done keeping in mind the available bandwidth resources. Hence, channel coding forms a very important preprocessing step in the transmission of digital data (bit stream). Since bandwidth is scarce and therefore expensive, a coding technique that requires fewer redundant bits without sacrificing error performance is highly desirable.

S. Faruque, *Free Space Laser Communication with Ambient Light Compensation*,
https://doi.org/10.1007/978-3-030-57484-0_1

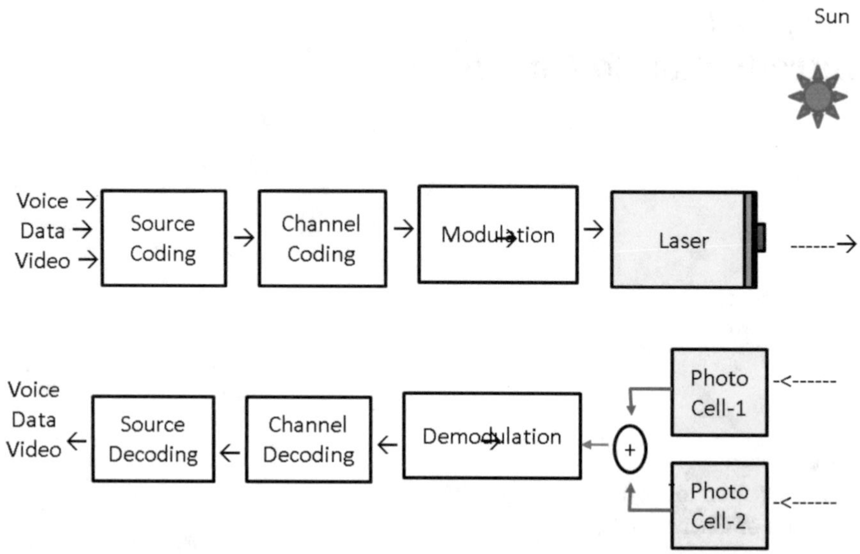

Sun

Fig. 1.1 Block diagram of a modern full-duplex communication system. The channel coding stage is shown as a dotted block

1.2   Types of Channel Coding

Channel coding attempts to utilize redundancy to minimize the effect of various channel impairments, such as noise and fading, and therefore increase the performance of the communication system. There are two basic ways of implementing redundancy to control errors. These are as follows:

- ARQ (automatic repeat request)
- FECC (forward error control coding)

1.2.1   ARQ Technique

The ARQ technique adds parity, or redundant bits, to the transmitted data stream that are used by the decoder to detect an error in the received data. When the receiver detects an error, it requests that the data be retransmitted by the receiver. This continues until the message is received correctly. In ARQ, the receiver does not attempt to correct the error, but rather it sends an alert to the transmitter in order to inform it that an error was detected and a retransmission is needed. This is known as a negative acknowledgement (NAK), and the transmitter retransmits the message upon receipt. If the message is error-free, the receiver sends an acknowledgement (ACK) to the transmitter. This form of error control is only capable of detecting

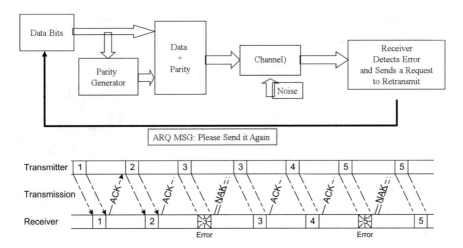

Fig. 1.2 Automatic repeat request (ARQ)

errors; it has no ability to correct errors that have been detected. This concept is presented in Fig. 1.2.

1.2.2   FECC Technique

In a system which utilizes FECC coding, the data are encoded with the redundant bits to allow the receiver not only to detect errors but also to correct them as well. In this system, a sequence of data signals is transformed into a longer sequence that contains enough redundancy to protect the data. This type of error control is also classified as channel coding, because these methods are often used to correct errors that are caused by channel noise. The goal of all FECC techniques is to detect and correct as many errors as possible without greatly increasing the data rate or the bandwidth. FECC codes are generally classified in two broad categories [4–10]:

- Block codes
- Convolutional codes
- Concatenated codes
- Orthogonal codes

Block Coding

In block coding, the information bits are segmented into blocks of k-data bits. The encoder transforms each k-bit data block into a larger block of n-bits, called coded information bits where $n > k$. The difference $(n - k)$ bits are the redundant bits, also known as parity bits. These redundant bits do not carry information, but enable the

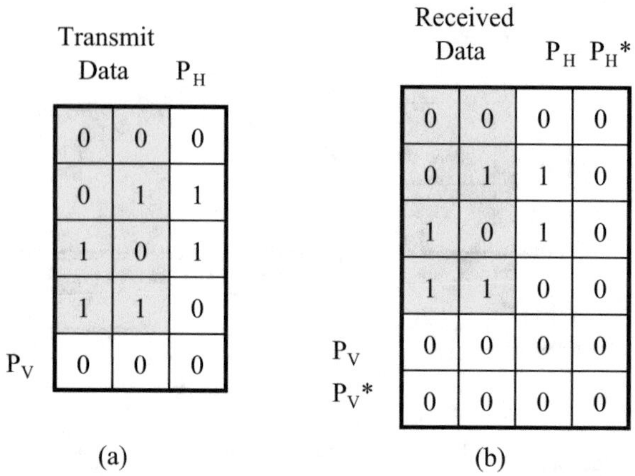

Fig. 1.3 Illustration of
(n, k) block code. (**a**)
Encoder, (**b**) decoder

Transmit
Data P_H

Received
Data P_H $P_H{}^*$

0	0	0
0	1	1
1	0	1
1	1	0

P_V | 0 | 0 | 0 |

0	0	0	0
0	1	1	0
1	0	1	0
1	1	0	0
0	0	0	0
0	0	0	0

P_V
$P_V{}^*$

(a)   (b)

detection and correction of errors. The code is referred to as (n, k) block code and the ratio k/n is known as "code rate."

Figure 1.3a shows an encoding scheme using (15, 8) block code where an 8-bit data block is formed as M-rows and N-columns ($M = 4$, $N = 2$). The product $MN = k = 8$ is the dimension of the information bits before coding. Next, a horizontal parity P_H is appended to each row and a vertical parity P_V is appended to each column. The resulting augmented dimension is given by the product $(M + 1)$ $(N + 1) = n = 15$, which is then transmitted to the receiver. The rate of this coding scheme is given by

$$r = \frac{k}{n} = \frac{MN}{(M + 1)(N + 1)} = \frac{8}{15}$$

Conversely, $1/r$ is a factor that increases the bit rate and hence the bandwidth. For example, if R_b is the bit rate before coding and r is the code rate, the coded bit rate will be R_b/r (b/s).

Upon receiving, the decoder (Fig. 1.3b) generates a new horizontal parity $P_H{}^*$ and a new vertical parity $P_V{}^*$. Now, if there is a single-bit error, there will be a parity check failure in the respective column and the respective row ($P_H{}^* = 1$ and $P_V{}^* = 1$), identifying the location of the error. Today, block coding is used in all digital communications.

Convolutional Coding

In convolutional coding:

- k information bits enter into the convolutional encoder sequentially,
- The convolutional encoder generates n parity bits ($n > k$).

Input Data (m)   Encoder   Encoded Output (U)

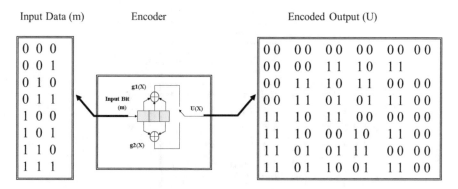

0 0 0				0 0	0 0	0 0	0 0	0 0	0 0
0 0 1				0 0	0 0	1 1	1 0	1 1	
0 1 0				0 0	1 1	1 0	1 1	0 0	0 0
0 1 1				0 0	1 1	0 1	0 1	1 1	0 0
1 0 0				1 1	1 0	1 1	0 0	0 0	0 0
1 0 1				1 1	1 0	0 0	1 0	1 1	0 0
1 1 0				1 1	0 1	0 1	1 1	0 0	0 0
1 1 1				1 1	0 1	1 0	0 1	1 1	0 0

Fig. 1.4 Encoder input/output relationships for $k = 3$, $r = 1/2$ (mapping)

- These parity bits (known as encoded bits) are modulated and transmitted through a channel.
- At the receive side, the receiver decodes by means of code correlation and regenerates the information bits.

As an illustration, a convolutional encoder is constructed with the following specifications:

- Constraint length $k = 3$
- Rate $r = 1/2$

The corresponding encoder, based on a 3-bit shift register and two exclusive OR gates, is shown in Fig. 1.4.

The operation of the encoder is as follows:

- The initial content of the encoder is 0 0 0.
- Three information bits enter into the 3-bit shift register sequentially, one bit at a time
- There are six shifts in the entire operation, generating six pairs of parity bits (12 parity bits) after which the shift register is cleared. The outcome is a rate $r = 6/12 = 1/2$ convolutional encoder.
- 3-bit data has $2^3 = 8$ combinations. Each combination generates a unique encoded bit pattern and stored in the lookup table.
- These encoded bits are modulated and transmitted through a channel.

Convolutional Decoder

Decoding is a process of code correlation, as presented below:

- A lookup table at the receiver contains the input/output bit sequences.
- In this process, the receiver compares the received data and generates a correlation value.
- The correlation value for each data set is stored in the lookup table.
- For $k = 3$, there are eight possible outputs.
- The receiver validates the received data pattern by means of code correlation.

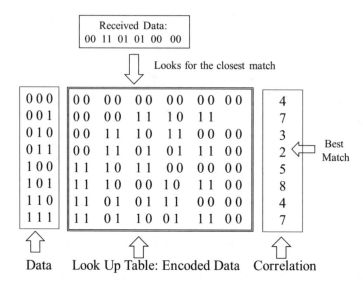

Fig. 1.5 Convolutional decoder. The received data is compared with the lookup table and finds the best match

- This is a process of finding the closest match, as shown in the table.
- The code rate is given by $r = \frac{1}{2}$ since there are 6 input bit sequences including the initial content of the shift register and there are 12 encoded data ($r = 6/12 = 1/2$) (Fig. 1.5).

Concatenated Coding

There are two types of concatenated coding:

- Series concatenated coding
- Parallel concatenated coding

Series concatenated coding, originally developed by Forney [5], is well-known for their excellent error control properties. A simple concatenated code, based on two codes in series, can be constructed as shown in Fig. 1.6.

Here, the high-speed user data is first encoded by means of an outer code, typically a block code. Next, the data and parity bits resulting from the outer code are interleaved and encoded by a rate 1/2 or rate 3/4 convolutional encoder inner code. The encoded bit stream is then modulated and transmitted through a channel.

On the receive side, the impaired code is first decoded by an inner decoder, typically a Viterbi decoder, and de-interleaved and then decoded by an outer decoder. The essential feature of the concatenated coding scheme is that any errors which do not get detected by the inner code are corrected by the outer code.

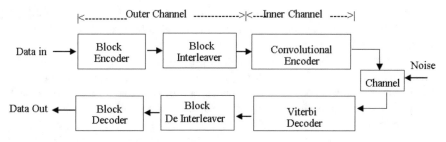

Fig. 1.6 Series concatenated coding based on block outer code and convolutional inner code

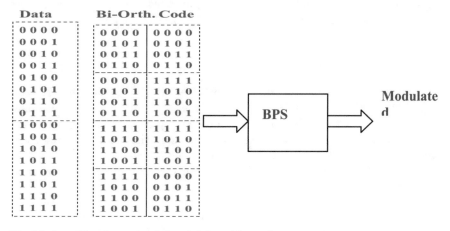

Fig. 1.7 Rate 1/2 orthogonal coded modulation with $n = 8$

Parallel concatenated coding, also known as turbo coding, uses two identical convolutional encoders, connected in parallel, and one internal interweaver. Turbo codes are a class of high-performance forward error correction (FEC) codes that closely approaches the theoretical channel capacity.

Orthogonal Coding

Orthogonal codes are binary valued and have equal number of 1s and 0s. These codes can be used as (n, k) block codes where a k-bit data set can be represented by a unique n-bit orthogonal code $(k < n)$. We illustrate this by means of an 8-bit orthogonal code, having 8-orthogonal and 8-antipodal codes for a total of 16 bi-orthogonal codes. We assume that an n-bit orthogonal code can be treated as an (n, k) block code. We now show that code rates such as rate 1/2, rate 3/4, and rate 1 are indeed available out of orthogonal codes. The principle is presented below:

A rate 1/2 orthogonal coded modulation with $n = 8$ can be constructed by inverse multiplexing the incoming traffic, R_b (b/s), into 4-parallel streams $(k = 4)$ as shown in Fig. 1.7. These bit streams, now reduced in speed to $R_b/4$ (b/s), are used to address

16 8-bit orthogonal codes, stored in an 8×16 ROM. The output of each ROM is a unique 8-bit orthogonal code, which is then modulated by a BPSK modulator and transmitted through a channel.

At the receiver, the incoming impaired orthogonal code is first examined by generating a parity bit. If the parity bit is 1, the received code is said to be in error. The impaired received code is then compared to a lookup table for a possible match. Once the closest approximation is achieved, the corresponding data is outputted from the lookup table. A brief description of the decoding principle is given below:

An n-bit orthogonal code has $n/2$ 1s and $n/2$ 0s; i.e., there are $n/2$ positions where 1 s and 0 s differ. Therefore, the distance between two orthogonal codes is $d = n/2$. This distance property can be used to detect an impaired received code by setting a threshold midway between two orthogonal codes. This is given by

$$d_{th} = \frac{n}{4}$$

where n is the code length and d_{th} is the threshold, which is midway between two valid orthogonal codes. Therefore, for the given 8-bit orthogonal code, we have $d_{th} = 8/4 = 2$. This mechanism offers a decision process, where the incoming impaired orthogonal code is examined for correlation with the neighboring codes for a possible match. Since the distance properties are the fundamental in error control coding, it can be shown that an n-bit orthogonal code can correct t errors, as given below:

$$t = \frac{n}{4} - 1$$

In the above equation, t is the number of errors that can be corrected by means of an n-bit orthogonal code. For example, a single error-correcting orthogonal code can be constructed by means of an 8-bit orthogonal code ($n = 8$). Similarly, a three-error-correcting, orthogonal code can be constructed by means of a 16-bit orthogonal code ($n = 16$) and so on. Table 1.1 shows a few orthogonal codes and their corresponding error-correcting capabilities.

Table 1.1 Orthogonal codes and the corresponding error-correcting capabilities

Code length n	Number of errors that can be corrected: $t = (n/4) - 1$
8	1
16	3
32	7
64	15
128	31

1.3 Design Considerations

In any communication system, the use of channel coding is often achieved at the expense of other system characteristics. Therefore, trade-offs often need to be made in order to develop a system that not only meets the performance needs but also adheres to the bandwidth and power constraints.

The first of these trade-offs is error performance versus bandwidth. Error-correction coding can be implemented to increase error performance, but these techniques require the transmission of additional bits, which will require an increase in bandwidth. Likewise, a system with limited power can reduce power without sacrificing error performance by implementing an FECC technique. This will again introduce an increase in the number of bits that needs to be transmitted by the system, again at the expense of bandwidth. Both these trade-offs assume a real-time communication system. However, if a non-real-time system is used, FECC coding can be used to improve performance and reduce power, but there will be an increase in delay instead of bandwidth. These trade-offs need to be considered when a communication system is being designed.

Several challenges, affecting minimum bandwidth and channel capacity, in designing digital communication systems, are introduced in this chapter. These are as follows [5, 10, 11]:

- Nyquist minimum bandwidth
- Shannon-Hartley capacity theorem
- Baseband modulation
- Waveform coding

1.3.1 Nyquist Minimum Bandwidth

Every realizable system that contains nonideal filtering will suffer from intersymbol interference (ISI). Intersymbol interference occurs when the tail of one pulse spills over and interferes with the correct detection of the adjacent symbol. Harry Nyquist showed in his 1928 paper "Certain Topics on Telegraph Transmission Theory" that the maximum theoretical number of symbols that can be received without ISI by a system with a transmission bandwidth of R_s Hertz is $R_s/2$.

Consequently, the sampling frequency, Ω_T, must be at least two times greater than the modulation frequency, Ω_m, of the transmitted data. This is known as the Nyquist condition and is written as

$$\Omega_T \geq 2\Omega_m$$

If the Nyquist condition is satisfied, the transmitted data can be fully reconstructed at the receiver with the use of an ideal low-pass filter with gain T and cutoff frequency Ω_c, so that

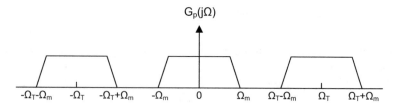

Fig. 1.8 Spectrum of the sampled signal with the Nyquist condition followed

Fig. 1.9 Spectrum of the sampled signal with the Nyquist condition not followed

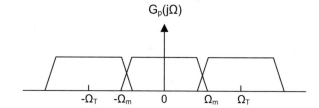

$$\Omega_m < \Omega_c < (\Omega_T - \Omega_m)$$

The spectrum of a signal, $g_p(t)$, sampled following the Nyquist condition is illustrated in Fig. 1.8. The spectrum of $g_p(t)$ sampled with $\Omega_T < 2\Omega_m$ is shown in Fig. 1.9.

The overlap in the sampled spectrum in Fig. 1.9 that is not present in Fig. 1.8 is known as aliasing and is the result of under-sampling. Aliasing leads to the receiver's inability to fully reconstruct the signal, because the spectrum $G_p(j\Omega)$ cannot be separated by filtering. Therefore, in order for a communication system to accurately and completely receive a transmitted signal, the modulation technique used needs to adhere to the Nyquist condition.

1.3.2 Shannon-Hartley Capacity Theorem

Any modern communication system strives to maximize bit rate while minimizing error probability, transmission energy, and bandwidth. Working against this goal is the presence of additive white Gaussian noise (AWGN) in all communication channels. Shannon showed in his 1948 paper that the system capacity C of a channel affected by AWGN can be written as [11]

$$C = W \log_2 \left(1 + \frac{S}{N}\right).$$

The system capacity is a function of bandwidth W, average received signal power S, and average noise power N. This relationship is known as the Shannon-Hartley

Fig. 1.10 Normalized
channel capacity versus
channel SNR

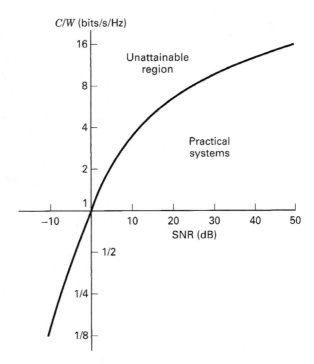

capacity theorem. When the bandwidth is in hertz and the base 2 logarithm is taken, the capacity is given in bits/s.

From this theorem, Shannon proved that it is theoretically possible to transmit information at a rate of R bits/s over a channel corrupted by AWGN with an arbitrarily small probability of error, so long as $R < C$. In order to accomplish this, a sufficiently complicated coding scheme needs to be implemented. It should be noted that Shannon's work sets a limit on channel capacity, not achievable error performance. Using the above equation, Shannon determined a bound for the achievable performance of a practical system. This bound is shown graphically in Fig. 1.10 as the normalized channel capacity C/W in bits/s/Hz versus the signal-to-noise ratio (SNR) of the channel (Fig. 1.11).

1.3.3 Baseband Modulation

In baseband modulation, the input waveforms are typically in the form of shaped pulses. Pulse and square waveforms are the most commonly used waveforms to represent digital data. Choice of return-to-zero (RZ) or non-return-to-zero (NRZ) data waveform depends on the application. NRZ is a binary code with no neutral rest condition and requires half the bandwidth required by the RZ data code. Also, it offers better noise immunity than the unipolar data waveforms like RZ data code.

Fig. 1.11 RZ and NRZ
binary data code

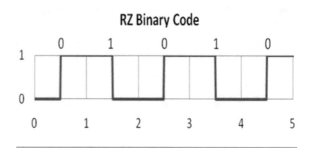

Fig. 1.12 Power spectrum
of RZ and NRZ binary data

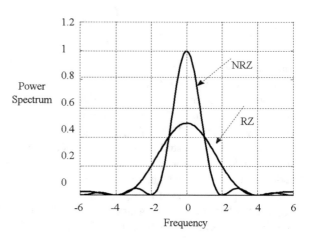

The bit durations for RZ and NRZ data are shown in Fig. 1.12. The transmission
bandwidth of NRZ and RZ data varies due to the fact that they have different bit
duration. As a result, the bandwidth associated with them also varies. Figure 1.12
shows the bandwidth and power density associated with both RZ and NRZ data.
According to the law of conservation of energy, the area under the two curves as
shown in Fig. 1.12 is the same. Therefore, the power magnitude $|P(\omega)|RZ$ is reduced
to half of $|P(\omega)|NRZ$.

1.3.4 Waveform Coding

Waveform coding is a form of channel coding where a set of waveforms is transformed into a set of orthogonal waveforms, so that the detection process is less subject to errors. There are two classes of waveform coding (a) M-ary signaling and (b) orthogonal coding. In M-ary signaling, a k-bit data set is used to address $M = 2^k$ modulated waveforms (e.g., MFSK). This process provides improved error performance at the expense of bandwidth. Similarly, in bi-orthogonal coding, a k-bit data set is directly mapped into $2n$ bi-orthogonal codes where n are the code lengths. This approach is also bandwidth inefficient, since the k-bit data set is directly mapped into 2×2^k bi-orthogonal codes.

In this book, we introduce an alternate method of waveform coding that does not consume additional bandwidth and offers protection against errors. In the proposed method, a high-speed data stream is inverse multiplexed into several parallel streams. These parallel streams, now reduced in speed, are grouped into a number of subsets and mapped into a predetermined group of bi-orthogonal codes and modulated by means of an MPSK modulator. This methodology substantially reduces the required number of waveforms and enhances transmission efficiency.

Figure 1.13 illustrates the concept. In Fig. 1.3a, we have a conventional method of waveform coding where a 4-bit data set is represented by $2^4 = 16$ waveforms. This scheme is commonly viewed as being bandwidth inefficient since we need 16 waveforms to transmit a 4-bit data.

Table 1.2 shows a comparison between the conventional M-ary signaling and the proposed method.

On the other hand, in the proposed method, as shown in Fig. 1.13b, when the same 4-bit data set is partitioned into two subsets, the number of waveforms reduces to 8. Similarly, in the conventional method, an 8-bit data set would require $2^8 = 256$ waveforms, while the proposed method requires only $2 \times 8 = 16$ waveforms. This is a substantial reduction of bandwidth indeed.

In the conventional method (Col-2, Table 1.2), a k-bit data set requires 2^k waveforms where $k = 1, 2, \ldots$. Thus the number of waveforms increases rapidly

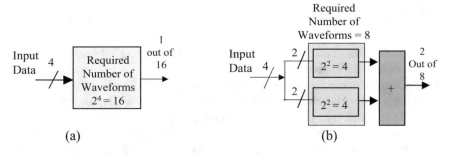

Fig. 1.13 Illustration of waveform coding. (**a**) Conventional method: a 4-bit data set requires 16 waveforms. (**b**) Proposed method: a 4-bit data set partitioned into two subsets requires only 8 waveforms

Table 1.2 Caption

# Bits (x)	Conventional method # Waveforms (2^x)	Proposed method # Waveforms (2x)	Bandwidth reduction Factor (2^x/2x)
1	2	2	1
2	4	4	1
3	8	6	1.333333333
4	16	8	2
5	32	10	3.2
6	64	12	5.333333333
7	128	14	9.142857143
8	256	16	16
9	512	18	28.44444444
10	1024	20	51.2
11	2048	22	93.09090909
12	4096	24	170.6666667
13	8192	26	315.0769231
14	16384	28	585.1428571
15	32768	30	1092.266667
16	65536	32	2048

as the length of the data set increases. For these reasons, the conventional method of waveform coding is bandwidth inefficient. In the proposed method, a k-bit data set requires only $2k$ waveforms where $k = 1, 2, \ldots$ (Col-3, Table 1.12). Clearly, the proposed method of waveform coding is bandwidth efficient. Our objective is to show that the proposed method of waveform coding applies to bi-orthogonal signaling. We also intend to show that there is a built-in error control mechanism in this scheme. Forward error control coding (FECC) schemes normally used in digital communication systems are not needed in the proposed method. Therefore, the proposed method is also cost-effective.

1.4 Conclusions

- Channel coding is a process of detecting and correcting bit errors in digital communication systems.
- It is also known as forward error control coding (FECC),
- Channel coding is performed both at the transmitter and at the receiver.
- At the transmit side, channel coding is referred to as encoder, where extra bits (parity bits) are added with the raw data before modulation.
- At the receive side, channel coding is referred to as the decoder. It enables the receiver to detect and correct errors, if they occur during transmission due to noise, interference, and fading.

This book presents the salient concepts, underlying principles, and practical realization of channel coding schemes currently used in digital communication system.

References

1. C.C. George et al., *Error Correction Coding for Digital Communications* (Plenum press, New York, 1981)
2. R.E. Blahut, *Theory and Practice of Error Control Codes* (Addison-Wesley, Reading, 1983)
3. Ungerboeck, Channel coding with multilevel/multiphase signals. IEEE Trans. Inf. Theory **IT28**, 55–67(1982, January)
4. S. Lin, D.J. Costello Jr., *Error Control Coding: Fundamentals and Applications* (Prentice-Hall, Inc., Englewood Cliffs, 1983)
5. G.D. Forney, Jr., *Concatenated Codes* (MIT Press, Cambridge, Mass., 1966). A.A. Azm, J. Kasparian, *Designing a Concatenated Coding System for Reliable Satellite and Space Communications*. Computers and Communications, 1997, Proceedings., Second IEEE Symposium on , vol. 1–3 Jul 1997, pp. 364–369
6. A.S. Tanenbaum, *Computer Networks* (Prentice Hall, Upper Saddle River, 2003), p. 223. ISBN 0-13-066102-3
7. P. Frenger, S. Parkvall, E. Dahlman, *Performance Comparison of HARQ with Chase Combining and Incremental Redundancy for HSDPA*. Vehicular Technology Conference, 2001. VTC 2001 Fall. IEEE VTS 54th 3. Piscataway Township, New Jersey: IEEE Operations Center. pp. 1829–1833 (October 2001). https://doi.org/10.1109/VTC.2001.956516. ISBN 0-7803-7005-8
8. I. Wegener, R.J. Pruim, *Complexity Theory*, ISBN 3-540-21045-8 (2005), p. 260
9. M. Furst, J. Saxe, M. Sipser, Parity, circuits, and the polynomial-time hierarchy. Math. Syst. Theory **17**(1), 13–27 (1984). https://doi.org/10.1007/BF01744431. Annu. Intl. Symp. Found. Computer Sci., 1981, Theory of Computing Systems
10. S. Faruque, *Bi-orthogonal Code Division Multiple Access System*, US Patent No. 198,719, Granted, March 6, 2001
11. C.E. Shannon, A mathematical theory of communication, Bell Syst. Tech. J. **27**, 379–423, 623–656, (1948, July, October)

Chapter 2
Automatic Repeat Request (ARQ)

Topics
- Introduction
- The Basic Concept of ARQ
- ARQ Building Blocks
- Construction of ARQ for Serial Data Processing
- Construction of ARQ for Parallel Data Processing
- Merits and Demerits of ARQ System
- Conclusions

2.1 Introduction to ARQ

Error control coding is a technique that adds redundant bits to minimize the effect of various channel impairments, such as noise and fading, and therefore increase the performance of the communication system [1–5]. There are two basic ways of implementing redundancy to control errors. The first is known as error detection and retransmission, which is also referred to as automatic repeat request (ARQ). The second method of error control through the use of redundancy is forward error control coding (FECC). In this chapter, our goal is to provide the basic understanding of error detection, focusing particularly on the following:

- Describe the basic functional blocks used in ARQ system
- Construct an ARQ system for processing serial data
- Construct an ARQ system for processing parallel data
- Provide examples

© The Editor(s) (if applicable) and The Author(s), under exclusive license to
Springer Nature Switzerland AG 2021
S. Faruque, *Free Space Laser Communication with Ambient Light Compensation*,
https://doi.org/10.1007/978-3-030-57484-0_2

2.2   The Basic Concept

The ARQ technique adds parity or redundant bits to the transmitted data stream that are used by the decoder to detect an error in the received data [6, 7]. When the receiver detects an error, it requests that the data be retransmitted by the receiver. This continues until the message is received correctly. In ARQ, the receiver does not attempt to correct the errors, but rather it sends an alert to the transmitter in order to inform it that an error was detected and a retransmission is needed. This is known as a negative acknowledgement, and the transmitter retransmits the message upon receipt. If the message is error free, the receiver sends an acknowledgement (ACK) to the transmitter. This form of error control is only capable of detecting errors; it has no ability to correct errors that have been detected. This concept is presented in Fig. 2.1.

Briefly, the operation can be described as follows:

- The transmitter generates "parity bits" from a block of raw data.
- The transmission includes both data and parity bits.
- The receiver computes the received data and looks for errors.
- If it detects an error, an ARQ message is sent over the reverse channel.
- Upon receiving the request, the transmitter retransmits the data.
- The process continues until the receiver declares a valid data.

Although ARQ system cannot correct errors, it is an important building block in non-real time digital communications where delay is not a problem, such as the Internet.

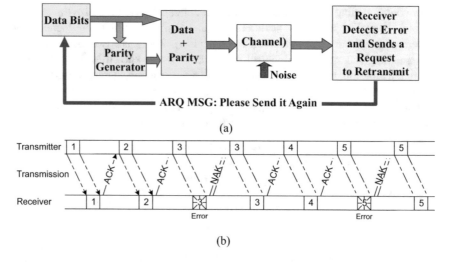

Fig. 2.1 Automatic repeat request (ARQ). (**a**) The functional block diagram. (**b**) Representation of negative acknowledgement (NAK) and positive acknowledgement (ACK)

2.3 ARQ Building Blocks

2.3.1 Parity

A parity bit, also known as a check bit, is a bit added to the end of a string of binary word that indicates whether the number of bits in the word with the value one is even or odd [8, 9]. There are two types of parity bits:

- Even parity bit
- Odd parity bit

Even Parity (Pe)
In the case of even parity, the number of bits whose value is 1 in a given word is counted. If the count of ones in a given word is even, the parity bit value is 0. This is defined as Pe= 0.

For example, a 2-bit word is said to be even, having an even parity bit as shown in the table below:

2-bit word	Even parity (Pe)
0 0	0
1 1	0

Similarly, a 3-bit word is said to be even, having an even parity bit as shown in the table below:

3-bit word	Even parity (Pe)
0 0 0	0
0 1 1	0
1 0 1	0
1 1 0	0

ODD Parity (Po)
In the case of odd parity, the number of bits whose value is 1 in a given word are counted. If the count of ones in a given word is odd, the parity bit value is 1. This is defined as Po= 1.

For example, a 2-bit word is said to be odd, having an odd parity bit as shown in the table below:

2-bit word	Odd parity (Po)
0 1	1
1 0	1

Similarly, a 3-bit word is said to be odd, having an odd parity bit as shown in the table below:

3-bit word	Odd parity (Po)
0 0 1	1
0 1 0	1
1 0 0	1
1 1 1	1

Likewise, a 4-bit data has $2^4 = 16$ words, having 8 even parities and 8 odd parities.

From the above examples, we see that, an n-bit word has $n/2$ even parity bits and $n/2$ odd parity bits.

2.3.2 Parity is an Arithmetic Operation

Parity is an arithmetic operation, also known as Modulo2 or MOD2 addition. The following examples illustrate the operation:

Even Parity
0 MOD2 Add 0 = 0 + 0 = 0
  1 MOD2 Add 1 = 1 + 1 = 0 (ignore the carry which is 1)

Odd Parity
0 MOD2 Add 1 = 0 + 1 = 1
  1 MOD2 Add 0 = 1 + 0 = 1

Similarly, a 3-bit word can be MOD2 added to generate an even parity and an odd parity as follows:

Even Parity
0 + 0 + 0 = 0
  0 + 1 + 1 = 0
  1 + 0 + 1 = 0
  1 + 1 + 0 = 0

Odd Parity
0 + 0 + 1 = 1
  0 + 1 + 0 = 1
  1 + 0 + 0 = 1
  1 + 1 + 1 = 1

Once again, in a given word, we see that when the number of 1's is even, the parity value is 0, and when the number 1's is odd, the parity value is 1. Therefore, by counting the number of 1's, the parity value of a given word can be determined simply by inspection.

2.3.3 Parity Generator

A parity generator is an array of exclusive OR (EXOR) gaits that generate parity bits known as odd parity or even parity. The parity generators are used in the transmit side as well as in the receive side. Briefly,

- ARQ is an arithmetic operation in digital system.
- It generates parity bits from a block of data.
- These parity bits are generated by means of a chain of EXCLUSIVE OR GATES (EXOR).

Our objective now is to examine exclusive OR gates and observe how parity bits can be generated by inspection.

Two Inputs EXOR

Consider the 2-input exclusive OR gate as shown in Fig. 2.2.
   The Boolean function of the exclusive OR gate is given by:

$$C = A\bar{B} + \bar{A}B$$
$$C = A \oplus B$$
$$C = A \ \text{EXOR} \ B$$

where

- $A = 0$ or 1
- $B = 0$ or 1
- $C =$ - output bit value

The truth table is given below:
Truth table of 2-Input Exor

A	B	C
0	0	0
0	1	1
1	0	1
1	1	0

Fig. 2.2 Two input exclusive OR gate

Fig. 2.3 Generation of parity bits by inspection

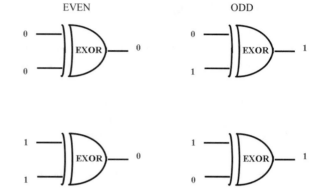

Fromm the above truth table, we see that:

- When both inputs are the same, the output is ZERO: even parity.
- When both inputs are not the same, the output is ONE: odd parity.

Therefore, we can also determine the parity simply by inspection as shown in Fig. 2.3.

2.3.4 Exclusive OR Chain Showing the Generation of Even Parity by Inspection

Figure 2.4 shows a chain of two exclusive OR gates. To determine the value of the parity, we use the following logic:

- If the input = even number of 1's, then the output is even: $P_e = 0$
- If the input = odd number of 1's, then the output is odd: $P_o = 1$

Therefore, by inspection, we find that:

- For the first EXOR chain: input = 1 1 0, which is even. Therefore, the parity value is 0, i.e., $P_e = 0$.
- For the second EXOR chain: input = 1 0 1, which is also even. Therefore, the parity value is 0, i.e., $P_e = 0$.

2.3.5 Exclusive OR Chain Showing the Generation of Odd Parity by Inspection

Figure 2.5 shows a chain of two exclusive OR gates. To determine the value of parity, we use the following logic:

Fig. 2.4 Exclusive OR chain showing the generation of even parity by inspection

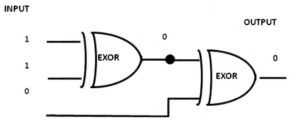

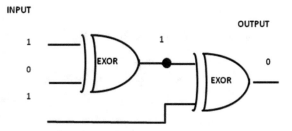

Fig. 2.5 Exclusive OR chain showing the generation of odd parity by inspection

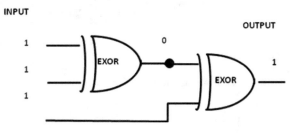

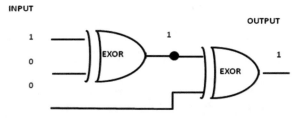

- If the input = even number of 1's, then the output is even: Pe = 0
- If the input = odd number of 1's, then the output is odd: Po = 1

  Therefore, by inspection, we find that:

- For the first EXOR chain: Input $= 1\ 1\ 1$, which is odd. Therefore, the parity value is 1, i.e., $Po = 1$
- For the second EXOR chain: Input $= 1\ 0\ 0$, which is also odd, Therefore, the parity value is 1, i.e., $Po = 1$.

Problem 2.1

Consider the exclusive OR chain as shown below:

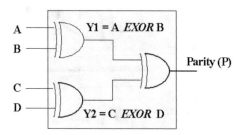

Find

(a) The following Boolean function:

- $Y1$
- $Y2$
- P(parity)

(b) If $A=1$, $B=0$, $C=1$, $D=0$, find the value of the corresponding parity value.
(c) If $A=1$, $B=1$, $C=0$, $D=1$, find the value of the corresponding parity value.
(d) Repeat part (b) and part (c) and give the parity values by inspection.

Solution
(a):

$Y1\quad = A \text{ EXOR } B$
$Y2\quad = C \text{ EXOR } D$
$P\quad\ \ = Y1 \text{ EXOR } Y2$
$\qquad = (A \text{ EXOR } B) \text{ EXOR } (C \text{ EXOR } D)$

(b) and (c):

The table below shows the parity values for the corresponding input bit values.

4-Bit Data Input   Parity

A	B	C	D	Y1	Y2	P
1	0	1	0	1	1	0
1	1	0	1	0	1	1

(d): Generation of parity by inspection

Since $A=1$, $B=0$, $C=1$, $D=0$
We can write:
$A+B+C+D=1+0+1+0=0$ (even parity)
Similarly, we have:
$A=1$, $B=1$, $C=0$, $D=1$
Therefore we can write:
$A+B+C+D=1+1+0+1=1$ (odd)
Figures below show the parity values obtained by inspection.

- The first parity generator yields an even parity since the data is even: $1+0+1+0=0$
- The second parity generator yields an Odd parity since the data is odd: $1+1+0+1=1$

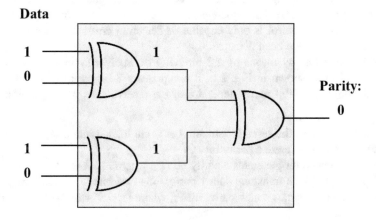

Data

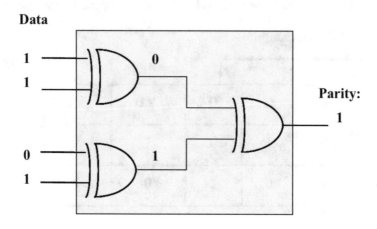

2.4   Construction of ARQ for Serial Data Processing

The ARQ technique adds a parity bit to the transmitted data stream that are used by the decoder to detect an error in the received data. Upon receiving, the receiver generates an additional parity bit out of the received data to detect an error and requests that the data be retransmitted, which is known as negative acknowledgement or "NAK," and the transmitter retransmits the message upon receipt. This continues until the message is received correctly. When the message is error free, the receiver sends a positive acknowledgement to the transmitter, known as "ACK." This form of error control is only capable of detecting errors; it has no ability to correct errors that have been detected.

We examine this by means of a 3-bit ARQ transceiver system for serial data communication as shown in Fig. 2.6. It comprises a 3-bit parity generator at the transmit side and a 4-bit parity generator at the receive side. The operation is as follows:

- At the transmitter, the serial bit stream A B C are loaded into a serial to parallel shift register to generate a parity bit P.
- The parity generator generates a parity bit (P), where $P = A+B+C = 1$ or 0.
- The transmitter then transmits data + parity $= 3+1 = 4$ bits to the receiver.
- The receiver computes a new parity bit Pr, where Pr $= P + A+B+C = 1$ or 0.

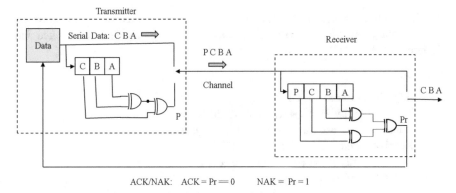

Fig. 2.6 Illustration of a 3-bit ARQ transceiver

- If Pr = 1, it declares an error, and an ARQ message is sent over the reverse channel as "NAK."
- Upon receiving the request, the transmitter retransmits the data.
- The process continues until the receiver declares a valid data by transmitting a 0 (Pr=0), which is designated as "ACK."

The ARQ process is governed by the following logic:

- If Pr=1, then retransmit the data (NAK).
- If Pr = 0, then validate the data (ACK).

where Pr is the parity generated by the receiver. This forms the basis of ARQ system.

Problem 2.2

Consider the 3-bit ARQ transceiver as shown below:

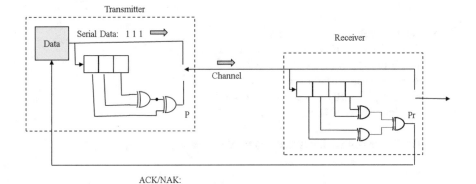

If the input bit pattern is 1 1 1, and if there are no transmission errors, show the flow of data throughout the ARQ transceiver. Give the parity values generated and declare a verdict (ACK or NACK).

Solution

- Input bit pattern = 1 1 1. (Odd)
- Transmit parity $P = 1 + 1 + 1 = 1$(odd parity)
- Transmit data = data + parity = 1 1 1 1 (even)
- Received data with no errors = 1 1 1 1 (even)
- Receive parity Pr = 1+ 1+1+1 = 0 (even)
- Verdict = ACK (no errors)

Figure below shows the data flow and parity.

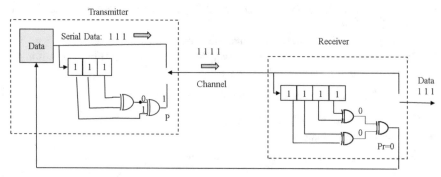

Pr = 0   Verdict =ACK

Problem 2.3
Consider the previous problem again. If the input bit pattern is 1 1 1, and if there is a transmission error for which the receiver receives a bit pattern 1 1 1 0, show the flow of data throughout the ARQ transceiver. Give the parity values generated and declare a verdict (ACK or NACK).

Solution

- Input bit pattern = 1 1 1. (odd)
- Transmit parity $P = 1 + 1 + 1 = 1$(odd parity)
- Transmit data = data + parity = 1 1 1 1 (even)
- Received data with one error = 1 1 1 0 (odd)
- Receive parity Pr = 1+1+1+0 = 1 (Odd)
- Verdict = NACK (error) – retransmit

The figure below shows the data flow and parity.

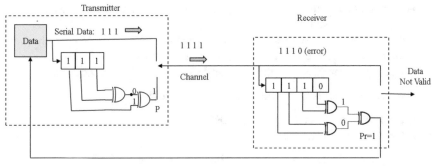

Pr = 1   Verdict = NACK (Retransmit)

2.5   Construction of ARQ for Parallel Data Processing

Figure 2.7 shows the functional diagram of a ARQ system supporting 4-bit parallel data communication between a transmitter and a receiver. It comprises a 4-bit parity generator at the transmit side and 5-bit parity generator at the receive side.

As shown in the figure:

- The transmitter generates a parity bit from a 4-bit word and transmits 4+1=5 bits to the receiver.

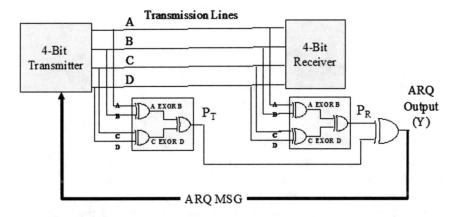

Fig. 2.7 ARQ for parallel data processing. This example shows that no error has been detected, and the verdict is ACK

- The receiver computes a new parity from the received data comprising data plus a parity bit and looks for 1 or 0.
- If it is 1, it declares an error, an ARQ message is sent over the reverse channel.
- Upon receiving the request, the transmitter retransmits the data.
- The process continues until the receiver declares a valid data by transmitting a 0.

The ARQ process is governed by the following logic:

- If $Y = 1$, then retransmit the data (NAK)
- If $Y = 0$, then validate the data (ACK)

To illustrate this, let's consider the following examples:

Example 1: (No Errors)

In this example (Fig. 2.7), let's assume that there is no error.

Input Data

$$A = 1$$
$$B = 0$$
$$C = 1$$
$$D = 0$$

Parity Bit Generated from 4-Bit Data:

$$P_T = A+B+C+D=1+0+1+0= 0$$

Transmit Bits

Data and Parity $= ABCDP_T = 10100$ (5 bits transmitted)

Receive Bits

Receive Bits:

$$A = 1$$
$$B = 0$$
$$C = 1$$
$$D = 0$$
$$P_T = 0$$

Receiver generates its own parity put of 5 bits as P_R:

- $P_R = A+B+C+D+P_T =1+0+1+0+0 = 0$ (Even)
- Verdict: No Error, ACK (Don't Care)

ARQ Example 2: With an Error

In this example, let's assume that there is a transmission error. Let this error be A, which is 0 instead of 1 (Fig. 2.8).

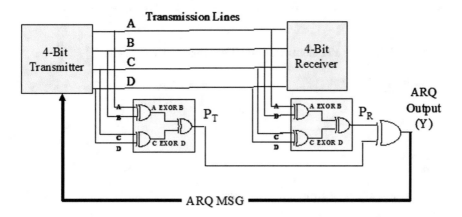

Fig. 2.8 ARQ for parallel data processing. This example shows that an error has been detected, and the verdict is NAK (retransmit data)

Transmit Bits

$A = 1$
$B = 0$
$C = 1$
$D = 0$
$P_T = 0$

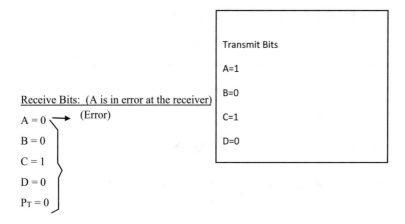

Receiver generates its own parity out of 5 bits as P_R:

- $P_R = A+B+C+D+P_T = 0+0+1+0+0 = 1$ (Odd)
- Verdict = NAK (retransmit)

Problem 2.4: [This Problem Assumes That the Parity is in Error]
Given:
Transmit Bits:

$A = 1$
$B = 0$
$C = 1$
$D = 0$
$P_T = 0$

Receive Bits: (P_T is in error at the receiver)

$A = 1$
$B = 0$
$C = 1$
$D = 0$
$P_T = 1$ $\longrightarrow$ (Parity is in Error)

Receiver generates its own parity out of 5 bits as P_R:

- $P_R = A+B+C+D+P_T= 1+0+1+0+1 = 1$ (Odd)
- Verdict = NAK (retransmit)

2.6 Merits and Demerits of ARQ System

This section will show that the ARQ system can only detect odd errors and cannot detect even errors. Let's examine this by means of examples.

2.6.1 Merits (ARQ Can Detect Odd Errors Only)

Consider the ARQ system as shown in Fig. 2.9. Here, we assume that there are three errors during transmission.

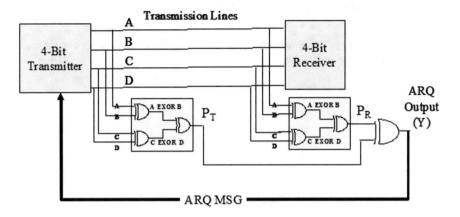

Fig. 2.9 A 4-bit ARQ for parallel data processing. This example will show that it can detect odd errors only

Transmit Bits

$A = 1$
$B = 0$
$C = 1$
$D = 0$
$P_T = 0$

Receive Bits: (There are three errors at the receiver)

$A = 1$
$B = 1$
$C = 1$ → There are three errors
$D = 1$
$P_T = 1$

Receiver generates its own parity out of 5 bits as P_R:

- $P_R = A+B+C+D+P_T= 1+1+111+1 = 1$ (Odd)
- Verdict = NAK (correct verdict)

Therefore, ARQ can detect odd errors.

2.6.2 Demerits (ARQ Cannot Detect Even Errors)

Consider the ARQ system as shown in Fig. 2.9 again. Here, we assume that there are two errors during transmission.

Transmit Bits

$A = 1$
$B = 0$
$C = 1$
$D = 0$
$P_T = 0$

Receive Bits: (There are two errors at the receiver)

$$A = 1$$
$$B = 1$$
$$C = 1 \quad \longrightarrow \quad \text{There are two errors}$$
$$D = 1$$
$$P_T = 0$$

Receiver generates its own parity out of 5 bits as P_R:

- $P_R = A+B+C+D+P_T= 1+1+1+1+0 = 0$ even)
- Verdict $=$ ACK (wrong verdict)

Therefore, ARQ cannot detect even errors.

2.7 Conclusions

- We have presented the basic concept of ARQ.
- We have also provided essential building blocks for ARQ.
- Construction of ARQ for serial data processing as well as for parallel data processing was shown to illustrate the concept.
- It is also shown that ARQ system cannot detect even errors and can detect multiple odd errors.

References

1. C.C. George et al., *Error Correction Coding for Digital Communications* (Plenum press, New York, 1981)
2. R.E. Blahut, *Theory and Practice of Error Control Codes* (Addison-Wesley, Reading, 1983)
3. Ungerboeck, Channel coding with multilevel/multiphase signals. IEEE Trans. Inf. Theory **IT28**, 55–67 (1982)
4. S. Lin, D.J. Costello Jr., *Error Control Coding: Fundamentals and Applications* (Prentice-Hall, Inc., Englewood Cliffs, 1983)
5. G.D. Forney, Jr., *Concatenated Codes* (MIT Press, Cambridge, Mass, 1966). A.A. Azm, L. Kasparian, Designing a concatenated coding system for reliable satellite and space communications. Computers and Communications, 1997, Proceedings, Second IEEE Symposium on vol. 1–3 Jul 1997, pp. 364–369
6. A.S. Tanenbaum, *Computer Networks* (Prentice Hall, Upper Saddle River, 2003), p. 223. ISBN 0-13-066102-3
7. P. Frenger, S. Parkvall, E. Dahlman, Performance comparison of HARQ with Chase combining and incremental redundancy for HSDPA. Vehicular Technology Conference, 2001. VTC 2001 Fall. IEEE VTS 54th 3. IEEE Operations Center, Piscataway Township, New Jersey, pp. 1829–1833, 2001, October. https://doi.org/10.1109/VTC.2001.956516. ISBN 0-7803-7005-8
8. I. Wegener, R.J. Pruim, *Complexity Theory*, 2005, p. 260. ISBN 3-540-21045-8
9. M. Furst, J. Saxe, M. Sipser, Parity, Circuits, and the Polynomial-Time Hierarchy. Annual International Symposium on Foundation of Computer Science, 1981, Theory of Computing Systems, vol. 17, no. 1, 1984, pp. 13–27. https://doi.org/10.1007/BF01744431

Chapter 3
Block Coding

Topics

3.1 Introduction to Block Coding

In FECC (forward error control coding), the data is encoded with the redundant bits to allow the receiver to not only detect errors but to correct them as well. In this system, a sequence of data signals is transformed into a longer sequence that contains enough redundancy to protect the data. This type of error control is also classified as channel coding because these methods are often used to correct errors that are caused by channel noise. The goal of all FECC techniques is to detect and correct as many errors as possible without greatly increasing the data rate or the bandwidth [1–5].

FECC codes are generally classified in two broad categories:

- Block codes
- Convolutional codes

Block codes are typically a memoryless technique that attempts to map the k input bits to n output bits, where $n > k$. The extra bits are referred to as parity bits. Block codes are usually denoted as (n, k) codes and have a code rate defined by k/n [6–8].

© The Editor(s) (if applicable) and The Author(s), under exclusive license to
Springer Nature Switzerland AG 2021
S. Faruque, *Free Space Laser Communication with Ambient Light Compensation*,
https://doi.org/10.1007/978-3-030-57484-0_3

Convolutional codes are a technique that uses memory to produce n output bits from k input bits. The code rate is again defined as k/n. The code rate is an indication of the amount of redundancy for a particular code. A low value for the code rate relates to more error-correcting ability but at the cost of increased bandwidth [9–14].

In any communication system, the use of channel coding is often achieved at the expense of other system characteristics. Therefore, trade-offs are often needed in order to develop a system that meets not only the performance needs but also adheres to the bandwidth and power constraints as well. The first of these trade-offs is error performance versus bandwidth. Error-correction coding can be implemented to increase error performance, but these techniques require the transmission of additional bits, which will require an increase in bandwidth. Likewise, a system with limited power can reduce power without sacrificing error performance by implementing an FECC technique. This will again introduce an increase in the number of bits that need to be transmitted by the system, again at the expense of bandwidth. Both these trade-offs assume a real-time communication system. However, if a non-real-time system is used, FECC coding can be used to improve performance and reduce power, but there will be an increase in delay instead of bandwidth. These trade-offs need to be considered when a communication system is being designed.

This chapter presents the key concepts, underlying principles, and practical application of block coding. Examples are provided to further illustrate the concept. In particular, the following topics are presented in this chapter:

- Block coding building blocks
- Typical rectangular block coding
- Code rate and bandwidth
- Modified rectangular block coding
- Modulation and transmission

3.2 Block Code Building Blocks

In block coding, the basic building block is a parity generator, where parity is an arithmetic operation, also known as Modulo2 or MOD2 addition. We have presented this topic in Chap. 2 in details. However, a short description is presented here for convenience.

A 2-bit word can be MOD2 added to generate an even parity and an odd parity as follows:

Even parity	ODD parity
$0 + 0 = 0$	$0 + 1 = 1$
$1 + 1 = 0$	$1 + 0 = 1$

Similarly, a 3-bit word can be MOD2 added to generate an even parity and an odd parity as follows:

Even parity	ODD parity
$0 + 0 + 0 = 0$	$0 + 0 + 1 = 1$
$0 + 1 + 1 = 0$	$0 + 1 + 0 = 1$
$1 + 0 + 1 = 0$	$1 + 0 + 0 = 1$
$1 + 1 + 0 = 0$	$1 + 1 + 1 = 1$

In a given word, we see that when the number of 1's is even, the parity value is 0, and when the number 1's is odd, the parity value is 1. Therefore, by counting the number of 1's, the parity value of a given word can be determined simply by inspection. We will use these analogies to construct block coding and see how block codes can detect and correct single errors.

3.3 Typical Rectangular Block Coding

3.3.1 Construction of Data Block

Block codes are a form of forward error correction (FECC) that can be used to both detect and correct errors. They are a type of parity check code that map k input binary bits to n output binary bits. They are characterized by the (n,k) notation. One type of block code is a rectangular code, which can be thought of as a parallel code structure.

In rectangular block coding, the k information bits are first segmented into rectangular blocks consisting of M rows and N columns. The input data stream is parsed into N-bit chunks and placed into the block one row at a time. Once the M rows are filled, the encoder adds parity, or redundant, bits to the block to form a larger block. This larger rectangular block consists of $(M + 1)$ rows and $(N + 1)$ columns and contains n coded information bits where $n > k$. The difference, $(n - k)$, are the parity bits. The purpose of the parity bits is to allow the decoder to detect and correct errors. The rate of the rectangular code is defined as:

$$r = \frac{k}{n} = \frac{MN}{(M + 1)(N + 1)}. \tag{3.1}$$

Once the information bits are placed into the rectangular block, a series of parity calculations are performed on the data. Modulo-2 addition, which is equivalent to the logical exclusive-or operation, is used to perform the calculations. The rules of modulo-2 addition are given in the previous section., which is given by the following equation:

$$0 + 0 = 0, 0 + 1 = 1, 1 + 0 = 1, 1 + 1 = 0, \tag{3.2}$$

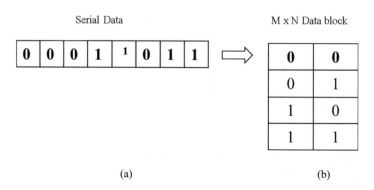

(a)   (b)

Fig. 3.1 (a) Serial data and (b) Rectangular data block having $M = 4$, $N = 2$ and $k = $ MN $= 8$

To illustrate the concept, an example of a (n, k) block coding scheme utilizing a (15, 8) rectangular block coding technique is displayed in Fig. 3.1a as an $M \times N$ matrix, where

- $M = $ number of rows $= 4$
- $N = $ number of columns $= 2$
- $n = (M + 1)(N + 1) = 15$
- $k = $ MN $= 8$
- $r = k/n = 8/15$ is the code rate

The input data stream is presented in Fig. 3.1a with the block assembled by the encoder given in Fig. 3.1b, which is the desired block of data having M rows and N columns, having $M = 4$, $N = 2$ and $k = $ MN $= 8$. The ratio k/n, defined as the code rate, is an important designed parameter in channel coding, where the inverse of code rate $1/r$ is a factor which expands the transmission bandwidth.

3.3.2   Encoder: Construction of Block Codes

Figure 3.2a shows the construction of a block code, commonly known as encoder. Here, the encoder performs a horizontal parity calculation on each row of data, and the result, P_H, is appended to the end of each row. Additionally, a vertical parity calculation is performed on each column of data with the result, P_V, being appended to the end of each column. An additional parity calculation is performed on the horizontal parity column, P_H, and placed at the end of the column. This ensures that both the parity row and parity column themselves have even parity. The entire block (data + parity) is then modulated and transmitted across the communication channel.

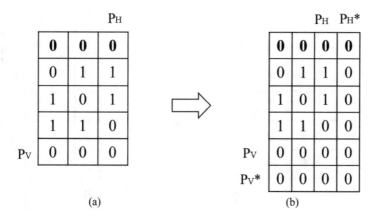

Fig. 3.2 (**a**) Encoder: Encoded data at the transmitter with vertical parity P_V and horizontal parity P_H. (**b**) Decoder: Received data with an additional vertical and horizontal parity P_V^* and horizontal parity P_H^*, respectively

3.3.3   Decoder: Detection and Correction of Errors

At the receiving end (Fig. 3.2b), the decoder performs a series of additional parity calculations on the received block. A new horizontal parity, P_H^*, is calculated with the result appended to the end of each row. A new vertical parity, P_V^*, is also calculated and placed at the end of each column. These additional parity calculations are utilized by the decoder in the error detection and correction process.

3.3.4   Example of Error Detection and Correction

Rectangular block coding is capable of detecting and correcting any single bit errors. If there is an error, a parity check failure ($P_H^* = 1$ and $P_V^* = 1$) will occur in the respective row and column. This allows the decoder to determine the location of the error and correct it. An example of an erroneous (15,8) rectangular block coding scheme is presented in Fig. 3.3.

The input data stream is the same as before. The encoded block of data, shown in Fig. 3.3a, is generated as follows:

At the Encoder
- Generate horizontal parity P_H from each row
- Generate vertical parity P_V from each column
- Transmit the entire content of the coded block to the receiver

At the Decoder
Let's assume that there is an error during transmission. The error is indicated in Fig. 3.3b by a circle. The decoder has no knowledge about this error. According to the protocol, the decoder performs the following:

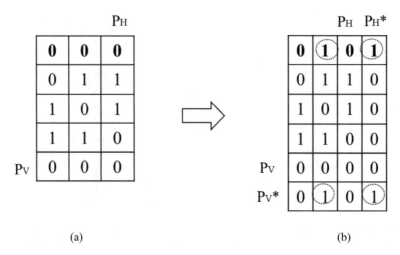

Fig. 3.3 Example of an erroneous (n,k) rectangular block code. (**a**) The encoder block. (**b**) The decoder block

- Decoder generates horizontal parity P_H^* from each row.
- Decoder generates vertical parity P_V^* from each column.

In Fig. 3.3b, it can be observed that the location of the error is determined by the row and column in which the parity check failures occur. The decoder would then flip this bit to correct the error. This type of FECC technique, however, is only capable of correcting single bit errors. Rectangular block codes can detect some multi-bit errors but are unable to correct them.

3.4   Code Rate and Bandwidth

3.4.1   Code Rate

In rectangular block coding, the k information bits are segmented into rectangular blocks consisting of M rows and N columns. Parity bits are generated from each row and each column to form a larger rectangular coded block. This larger rectangular block consists of $(M + 1)$ rows and $(N + 1)$ columns and contains n coded information bits where $n > k$. The extra bits, $(n-k)$, are referred to as parity bits. The purpose of the parity bits is to allow the decoder to detect and correct errors. Block codes are usually denoted as (n, k) codes and have a code rate defined by k/n. This is given by the following equation:

$$r = k/n \qquad\qquad (3.3)$$

Where,

$r =$ Code rate
$k =$ Number of Uncoded bit
$n =$ /Number of coded bits

3.4.2 Bandwidth

In block coding, redundant bits (parity bits) are transmitted along with information bits, which will require an increase in bandwidth. Before modulation, this is given by:

$$\text{Bandwidth} : \text{BW} = R_b/r = R_b(n/k) \tag{3.4}$$

where

$R_b =$ input bit rate (b/s)
$n =$ number of bits after coding
$k =$ number of bits before coding

Problem 3.1
A rectangular block code is constructed by using M rows and N columns, where $M = 4$ and $N = 3$. Calculate the code rate.

Solution:

$$\text{Code Rate } r = \frac{k}{n} = \frac{MN}{(M+1)(N+1)} = (4 \times 3)/(5 \times 4) = 12/20 = 0.6$$

Problem 3.2
Consider the block of data as shown below:

Data

0	0	0	0
0	1	0	1
0	0	1	1
0	1	1	0

(a) Construct the encoded data block at the transmitter.
(b) If there is no error, construct the decoded data block at the receiver.

(c) If the bit in location row 1 and column 1 is in error, show how the receiver
   detects and corrects the error.

Solution:

(a) Encoded Data Block

   At the transmit end, the encoder performs a series of parity calculations on the
data block. A horizontal parity, P_H, is calculated with the result appended to the
end of each row. A vertical parity, P_V, is also calculated and placed at the end of
each column. The entire block is then transmitted to the receiver.

		Data			P_H
	0	0	0	0	0
	0	1	0	1	0
	0	0	1	1	0
	0	1	1	0	0
P_V	0	0	0	0	0

(b) Decoded Data Block (No Error)

   At the receiving end, the decoder performs a series of additional parity calculations on the received block. A new horizontal parity, $P_H{}^*$, is calculated with the
result appended to the end of each row. A new vertical parity, $P_V{}^*$, is also calculated
and placed at the end of each column. These additional parity values are all zeros.
Therefore, there are no errors.

		Data			P_H	$P_H{}^*$
	0	0	0	0	0	0
	0	1	0	1	0	0
	0	0	1	1	0	0
	0	1	1	0	0	0
P_V	0	0	0	0	0	0
$P_V{}^*$	0	0	0	0	0	0

(c) Decoded Data Block (With an Error)

Since the error occurs in location row 1 and column 1 (dotted circle) the corresponding horizontal and vertical parity values are changed. Indicating the location of the error. Therefore, this bit can be inverted to make the correction. See figure below:

Error

	Data				PH	PH*
①	0	0	0	0	1	
0	1	0	1	0	0	
0	0	1	1	0	0	
0	1	1	0	0	0	
PV	0	0	0	0	0	0
PV*	1	0	0	0	0	0

Problem 3.3
Given:

- Bit rate $R_b = 10$ kb/s
- (49×36) block code

Find:

(a) The (49×36) block coding scheme
(b) Calculate the code rate r
(c) Calculate the bandwidth without modulation
(d) If the input bit rate $R_b = 10$ kb/s, calculate the required transmission bandwidth

Solution:

(a) (49×36) block coding scheme:

- $k = 36$. Therefore the uncoded block can be constructed as: $M \times N = 6 \times 6$ matrix
- $n = 49$. Therefore the coded block can be constructed as: $(M + 1)(N + 1) = 7 \times 7$ matrix

(b) Code rate $r = k/n = 36/49 = 0.734694$

(c) Bw $= R_b/r =$ 10 kb/0.734694 $=$ 13.61111 kHz

Drill Exercise
Given:

- A 16-bit data block, arranged as a 4 × 4 matrix as shown below.
- During transmission, the bit in location row 3 and column 3 gets corrupted, and the receiver decodes it as "0" as indicated by a circle.

(a) Construct the encoded data block.
(b) Construct the decoded data block and show how the receiver detects and corrects the error bit.

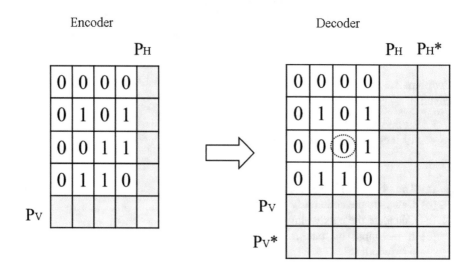

3.5   Modified Rectangular Block Coding

3.5.1   Encoder

In an attempt to enhance the error correcting capability, a modified technique was developed [15]. This modified block coding scheme adds fewer parity bits, which results in a saving of bandwidth. As the input data stream enters the encoder, it is parsed into smaller k-bit data chunks. A horizontal parity calculation is then performed on the k-bit data chunk. The parity calculations are computed using modulo-2 addition. The k-bit data chunk is then placed into one of two rectangular blocks, each containing M rows and k columns. If the result of the horizontal parity calculation is even (a result of 0), the k-bit data chunk is placed into the "even" block. If the result of the horizontal parity calculation is odd (a result of 1), the k-bit data chunk is placed into the "odd" block. This technique eliminates the need to transmit the horizontal parity column, P_H, required in typical rectangular block coding. This is possible because the horizontal parity of the blocks is known to the decoder, since only k-bit data chunks consisting of even (or odd) parity are present in each block. The savings depend on the number of rows in each block used by the encoder. The modified scheme then performs a vertical parity calculation, P_V, and appends it to the end of each column. This is illustrated in Fig. 3.4, using 2-bit data chunks and the same 8-bit input data stream used in the rectangular block coding examples and shown previously.

3.5.2   Decoder

As the data are received, they are placed in the appropriate parity block, and a timestamp is appended to the end of the row corresponding to the order of arrival. This does not add to the amount of transmitted data because it is performed at the decoder. Once the parity row has been received, the decoder calculates a new vertical parity, $P_V{}^*$, for each column. It also calculates a new horizontal parity, $P_H{}^*$, for each row. For the horizontal parity calculations, the encoder uses the received data along with the parity value of the block. This is the same method as that used in typical rectangular block coding.

The modified technique can detect and correct any single-bit error in each block for a total of two errors. The location of the error can be determined from the row and column in which a parity check failure ($P_H{}^* = 1$ and $P_V{}^* = 1$) occurred. A visual

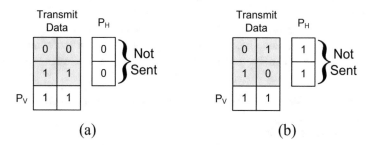

(a)   (b)

Fig. 3.4 Example encoder using the modified block coding technique. (**a**) Even parity block. (**b**) Odd parity block

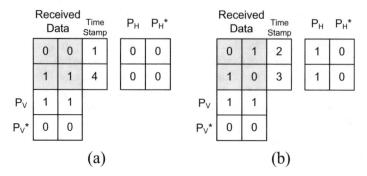

Fig. 3.5 Example decoder using the modified block coding technique. (**a**) Even parity block. (**b**) Odd parity block

example of a decoder using the modified block coding technique is displayed in Fig. 3.5.

Not only does the modified technique require fewer parity bits than typical rectangular block coding, but it also results in greater coding strength. This is due to the fact that there are fewer bits being transmitted over a potentially noisy channel. In typical rectangular block coding, the horizontal parity column is transmitted, which makes it susceptible to corruption by channel noise. If the horizontal parity bits are corrupted, they will affect the ability of the decoder to detect and correct errors in the received data. The modified technique proposed does not suffer from this because the horizontal parity column for each block is not transmitted over the communication channel. Instead, it is determined by the decoder, based on the received demodulation frequency. Moreover, the proposed modified technique is capable of correcting any single-bit error per block, for a total of two errors.

A major difference between the modified block coding scheme and typical rectangular block coding is the method in which the data are modulated and transmitted across the communication channel. If the entire block was modulated and transmitted as in rectangular block coding, it would not be possible to reassemble the data. This is because the decoder would not know the order in which the data entered the encoder without the encoder adding a timestamp to the data. The timestamp would eliminate the bandwidth saved by the modified technique.

To accommodate this, as each parsed chunk of the input data stream is placed into one of the blocks, it is also modulated and transmitted across the communication channel. Modulation can be performed using any of the following modulation schemes:

- Amplitude shift keying (ASK), also known as on-off-keying (OOK)
- Frequency shift keying (FSK)
- Phase shift keying (PSK).

Once the parity row of the block has been calculated, it is also modulated using the appropriate modulator and transmitted by means of the respective carrier frequency. Since there are two parity blocks, the modified technique corrects two errors at the expense of two carrier frequencies (one for each parity block).

XXXXXXXXXXXXXXXXXXXXXXXXXXXXXXXXXXXXXXXXXXXXXXXXXX

XXXXXXXXXXXXXXXXXXXXXXXXXXXXXXXXXXXXX

XXXXXXXXXXXXXXXXXX

3.6 Modulation and Transmission at a Glance

Once the data is encoded, it needs to be modulated before transmission. This is a fundamental requirement in wireless communication, where modulation is a technique that changes the characteristics of the carrier frequency in accordance to the input digital signal. Furthermore, the radiating device is an antenna, which is a reciprocal device that transmits and receives sinusoidal waves. The size of the antenna depends on the wavelength (λ) of the sinusoidal wave where,

$$\lambda = c/f \text{ meter}$$

c = velocity of light = 3×10^8 m/s
f = frequency of the sinusoidal wave, also known as "carrier frequency"

Therefore, a carrier frequency much higher than the input signal is required to keep the size of the antenna at an acceptable limit. For these reasons, a high-frequency carrier signal is used in the modulation process. In this process, the low-frequency input signal changes the characteristics of the high-frequency sinusoidal waveform in a certain manner, depending on the modulation technique. For digital signals, there are several modulation techniques available. The three main digital modulation techniques are:

- Amplitude shift keying (ASK)
- Frequency shift keying (FSK).
- Phase shift keying (PSK).

3.6.1 Amplitude Shift Keying (ASK) Modulation

Amplitude shift keying (ASK), also known as on-off-keying (OOK),is a method of digital modulation that utilizes amplitude shifting of the relative amplitude of the career frequency [16–17]. The signal to be modulated and transmitted is binary; this is referred to as ASK, where the amplitude of the carrier changes in discrete levels, in accordance to the input signal.

Figure 3.6 shows a functional diagram of a typical ASK modulator for different input bit sequences, where

- Input digital signal is the information we want to transmit.
- Carrier is the radio frequency without modulation.
- Output is the ASK modulated carrier, which has two amplitudes corresponding to the binary input signal. For binary signal 1, the carrier is ON. For the binary signal 0, the carrier is Off; however, a small residual signal may remain due to noise, interference, etc.

As shown in Fig. 3.6, the amplitude of the carrier changes in discrete levels, in accordance to the input signal, where,

Input data: $m(t) = 0$ or 1
Carrier frequency: $C(t) = A \, Cos(\omega t)$
Modulated carrier: $S(t) = m(t)C(t) = m(t)A \, Cos(\omega t)$

   Therefore,

For $m(t) = 1$: $S(t) = A \, Cos(\omega t)$, i.e., the carrier is ON
For $m(t) = 0$: $S(t) = 0$, i.e., the carrier is OFF

where A is the amplitude and ω is the frequency of the carrier.

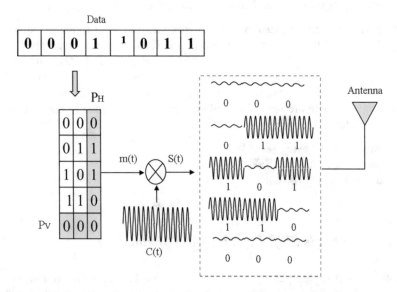

Fig. 3.6 Amplitude shift keying (ASK), also known as on-off-keying (OOK). The input encoded data block is transmitted row by row. The amplitude of the carrier frequency changes in accordance to the input digital signal

3.6.2   Amplitude Shift Keying (ASK) Demodulation

Once the modulated binary data has been transmitted, it needs to be received and demodulated. This is often accomplished with the use of a bandpass filter. In the case of ASK, the receiver needs to utilize one bandpass filter that is tuned to the appropriate carrier frequency. As the signal enters the receiver, it passes through the filter, and a decision as to the value of each bit is made to recover the encoded data block, along with horizontal and vertical parities. Next, the receiver appends horizontal and vertical parities P_H^* and P_V^* to check parity failures and recovers the data block. This is shown in Fig. 3.7 having no errors. If there is an error, there will be a parity failure in P_H^* and P_V^* to pin point the error.

XXXXXXXXXXXXXXXXXXXXXXXXXXXXXXXXXXXXXXXXXXXXX

xxxxxxxxxxxxxxxxxxxxxxxxxxxxxxxxxx

xxxxxxxxxxxxxxx

3.6.3   Frequency Shift Keying (FSK) Modulation

Frequency shift keying (FSK) is a method of digital modulation that utilizes frequency shifting of the relative frequency content of the signal [16–17]. The signal to be modulated and transmitted is binary; this is referred to as binary FSK (BFSK),

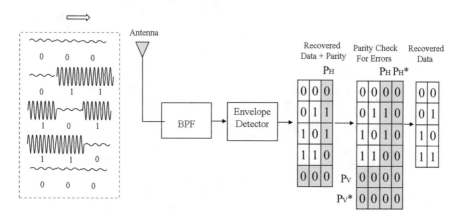

Fig. 3.7 Data recovery process in ASK, showing no errors. If there is an error, there will be a parity failure in P_H^* and P_V^* to pinpoint the error

where the carrier frequency changes in discrete levels, in accordance with the input signal.

Figure 3.8 shows a functional diagram of a typical FSK modulator for different input bit sequences, where

- Input digital signal is the information we want to transmit.
- Carrier is the radio frequency without modulation.
- Output is the FSK modulated carrier, which has two frequencies ω_1 and ω_2, corresponding to the binary input signal.
- These frequencies correspond to the messages binary 0 and 1, respectively.

As shown in Fig. 3.8, the frequency of the carrier changes in discrete levels, in accordance to the input signals. We have:

Input data: $m(t) = 0$ or 1
Carrier frequency: $C(t) = A \cos (\omega t)$
Modulated carrier: $S(t) = A \cos(\omega - \Delta\omega)t$, For $m(t)=1$
                              $S(t) = A \cos(\omega + \Delta\omega)t$, For $m(t) = 0$

where

- A = amplitude of the carrier
- ω = nominal frequency of the carrier frequency
- $\Delta\omega$ = frequency deviation

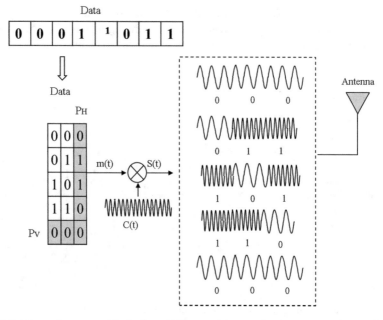

Fig. 3.8 Binary frequency shift keying (BFSK) modulation. The input encoded data block is transmitted row by row. The frequency of the carrier changes in accordance to the input digital signal

3.6.4   Frequency Shift Keying (FSK) Demodulation

Once the modulated binary data has been transmitted, it needs to be received and demodulated. This is often accomplished with the use of bandpass filters. In the case of binary FSK, the receiver needs to utilize two bandpass filters that are tuned to the appropriate frequencies. Since the nominal carrier frequency and the frequency deviation are known, this is relatively straightforward. One bandpass filter will be centered at the frequency ω_1, and the other at ω_2. As the signal enters the receiver, it passes through the filters, and a decision as to the value of each bit is made. This is shown in Fig. 3.9. In order to assure that the bits are decoded correctly, the frequency deviation needs to be chosen with the limitations of the filters in mind to eliminate crossover.

3.6.5   Phase Shift Keying (PSK) Modulation

Phase shift keying (PSK) is a method of digital modulation that utilizes phase of the carrier to represent digital signal [16–17]. The signal to be modulated and transmitted is binary; this is referred to as binary PSK (BPSK), where the phase of the carrier changes in discrete levels, in accordance with the input signal as shown below:

- Binary 0 (Bit 0): $Phase_1 = 0$ deg.
- Binary 1 (Bit 1): $Phase_2 = 180$ deg.

Figure 3.10 shows a functional diagram of a typical binary phase shift keying (BPSK) modulator for different input bit sequences, where

- Input digital signal is the information we want to transmit.

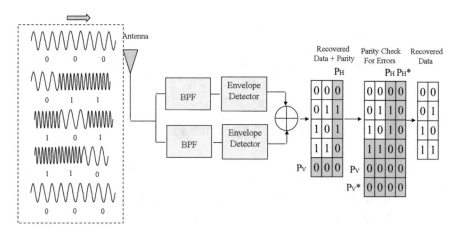

Fig. 3.9 Binary FSK detector utilizing two matched bandpass filters

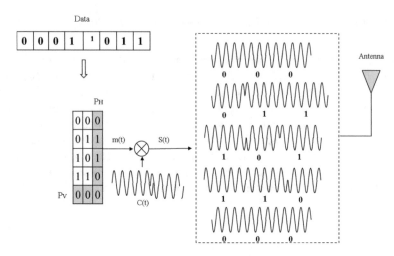

Fig. 3.10 Binary phase shift keying (BPSK) modulation. The input encoded data block is transmitted row by row. The phase of the carrier frequency changes in accordance to the input digital signal

- Carrier is the radio frequency without modulation.
- Output is the BPSK modulated carrier, which has two phases φ_1 and φ_2 corresponding to the two information bits.

As shown in Fig. 3.10, the phase of the carrier changes in discrete levels, in accordance to the input signal. We have:

Input data: $m(t) = 0$ or 1
Carrier frequency: $C(t) = A \, \mathrm{Cos}\,(\omega t)$
Modulated carrier: $S(t) = A \, \mathrm{Cos}(\omega+\varphi)t$

 where

- $\varphi) = 0°$, $m(t)$ 0
- $\varphi) = 180°$, $m(t)$ 1
- $A =$ amplitude of the carrier
- $\omega =$ frequency of the carrier frequency

3.6.6   Phase Shift Keying (PSK) Demodulation

Once the modulated binary data has been transmitted, it needs to be received and demodulated. This is often accomplished with the use of a phase detector, typically known as phase-locked loop (PLL). As the signal enters the receiver, it passes through the PLL. The PLL locks to the incoming carrier frequency and tracks the

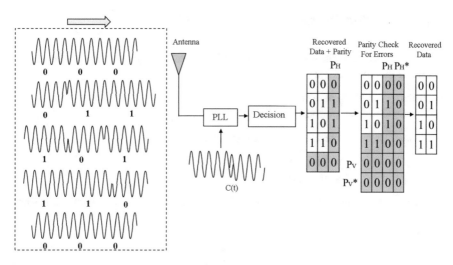

Fig. 3.11 Binary PSK detector showing data recovery process

variations in frequency and phase. This is known as coherent detection technique, where the knowledge of the carrier frequency and phase must be known to the receiver. Figure 3.11 shows a simplified diagram of a BPSK demodulator along with the data recovery process. In order to assure that the bits are decoded correctly, the phase deviation needs to be chosen with the limitations of the PLL in mind to eliminate crossover.

XXXXXXXXXXXXXXXXXXXXXXXXXXXXXXXXXXXXXXXXXX

XXXXXXXXXXXXXXXXXXXXXXXXXXXXXXXXXX

3.7   Estimation of Transmission Bandwidth

In wireless communications, the scarcity of RF spectrum is well-known. For this reason, we have to be vigilant about using transmission bandwidth in error control coding and modulation. The transmission bandwidth depends on:

- Spectral response of the encoded data
- Spectral response of the carrier frequency
- Modulation type (ASK, FSK, PSK), etc.

3.7.1 Spectral Response of the Encoded Data

In digital communications, data is generally referred to as a nonperiodic digital signal. It has two values:

- Binary-1 = high, period = T
- Binary-0 = low, period = T

Also, data can be represented in two ways:

- Time domain representation
- Frequency domain representation

The time domain representation (Fig. 3.12a), known as non-return-to-zero (NRZ), is given by:

$$V(t) = V \quad < 0 < t < T$$
$$= 0 \quad \text{elsewhere} \tag{3.5}$$

The frequency domain representation is given by: "Fourier transform":

$$V(\omega) = \int_0^T V.e^{-j\omega t} dt \tag{3.6}$$

$$|V(\omega)| = VT \left[\frac{Sin(\omega T/2)}{\omega T/2} \right] \tag{3.7}$$

$$P(\omega) = \left(\frac{1}{T} \right) |V(\omega)|^2 = V^2 T \left[\frac{Sin(\omega T/2)}{\omega T/2} \right]^2$$

Here, $P(\omega)$ is the power spectral density. This is plotted in Fig. 3.12b. The main lobe corresponds to the fundamental frequency, and side lobes correspond to harmonic components. The bandwidth of the power spectrum is proportional to the frequency. In practice, the side lobes are filtered out since they are relatively insignificant with respect to the main lobe. Therefore, the one-sided bandwidth is given by the ratio $f/f_b = 1$. In other words, the one-sided bandwidth $= f = f_b$, where $f_b = R_b = 1/T$, T being the bit duration.

The general equation for two-sided response is given by:

$$V(\omega) = \int_{-\infty}^{\infty} V(t).e^{-j\omega t} dt$$

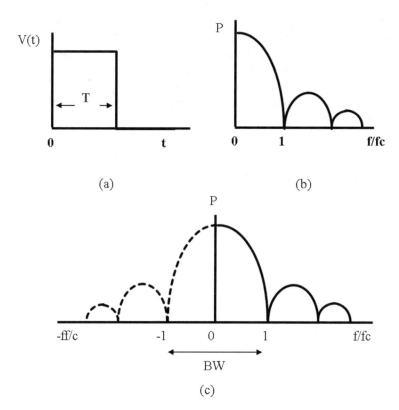

Fig. 3.12 (a) Discrete time digital signal (b) its one-sided power spectral density and (c) two-sided power spectral density. The bandwidth associated with the non-return-to-zero (NRz) data is $2R_b$, where R_b is the bit rate

In this case, $V(\omega)$ is called two-sided spectrum of $V(t)$. This is due to both positive and negative frequencies used in the integral. The function can be a voltage or a current Fig. 3.12c shows the two sided response, where the bandwidth is determined by the main lobe as shown below:

$$\text{Two sided bandwidth (BW)} = 2R_b(R_b = \text{Bit rate before coding}) \qquad (3.8)$$

Important Note

1. If R_b is the bit rate before coding, and if the data is NRZ, then the bandwidth associated with the raw data will be $2R_b$. For example, if the bit rate before coding is 10 kb/s, then the bandwidth associated with the raw data will be 2×10 kb/s $= 20$ kHz.

2. If R_b is the bit rate before coding, code rate is r, and if the data is NRZ, then the bit rate after coding will be R_b(coded) $=$ R_b(uncoded)r. The corresponding bandwidth associated with the coded data will be $2R_b$ (coded) $=$ $2R_b$ (uncoded/r. For example, if the bit rate before coding is 10 kb/s and the code rate $r =$ 1/2, the coded bit rate will be R_b (coded) $=$ R_b (uncoded)/$r =$ 10/0.5 $=$ 20 kb/s. The corresponding bandwidth associated with the coded data will be $2 \times 20 = 40$ kHz.

3.7.2 Spectral Response of the Carrier Frequency Before Modulation

A carrier frequency is essentially a sinusoidal waveform, which is periodic and continuous with respect to time. It has one frequency component. For example, the sine wave is described by the following time domain equation:

$$V(t) = V_p \mathrm{Sin}(\omega t_c) \qquad (3.9)$$

where

$V_p =$ peak voltage

- $\omega_c = 2\pi f_c$
- $f_c =$ carrier frequency in Hz

Figure 3.13 shows the characteristics of a sine wave and its spectral response. Since the frequency is constant, its spectral response is located in the horizontal axis, and the peak voltage is shown in the vertical axis. The corresponding bandwidth is zero.

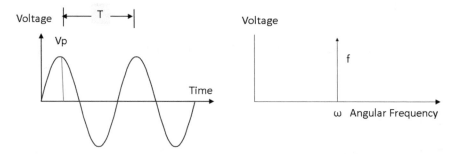

Fig. 3.13 A sine wave and its frequency response

3.7.3 ASK Bandwidth at a Glance

In ASK, the amplitude of the carriers changes in discrete levels, in accordance with the input signal, where

Input data: $m(t) = 0$ or 1
Carrier frequency: $C(t) = A_c \cos(\omega_c t)$
Modulated carrier: $S(t) = m(t)C(t) = m(t)A_c \cos(\omega_c t)$

Since $m(t)$ is the input digital signal, and it contains an infinite number of harmonically related sinusoidal waveforms and that we keep the fundamental and filter out the higher-order components, we write:

$$m(t) = A_m \sin(\omega_m t)$$

The ASK modulated signal then becomes:

$$S(t) = m(t)S(t) = A_m A_c \sin(\omega_m t) \cos(\omega_c)$$
$$= A_m A_c \cos(\omega_c + \omega_m)$$

The spectral response is depicted in Fig. 3.14. Notice that the spectral response after ASK modulation is the shifted version of the NRZ data. Bandwidth is given by: $\text{BW} = 2R_b$ (coded), where R_b is the coded bit rate.

XXXXXXXXXXXXXXXXXXXXXXXXXXXXXXXXXX

XXXXXXXXXXXXXXXXXXXXX

3.7.4 FSK Bandwidth at a Glance

In FSK, the frequency of the carrier changes in two discrete levels, in accordance to the input signals. We have:

Input data: $m(t) = 0$ or 1
Carrier frequency: $C(t) = A \cos(\omega t)$
Modulated carrier: $S(t) = A \cos(\omega - \Delta\omega)t$, For $m(t) = 1$
                      $S(t) = A \cos(\omega + \Delta\omega)t$, For $m(t) = 0$

where

- $S(t) = $ the modulated carrier
- $A = $ amplitude of the carrier
- $\omega = $ nominal frequency of the carrier frequency
- $\Delta\omega = $ frequency deviation

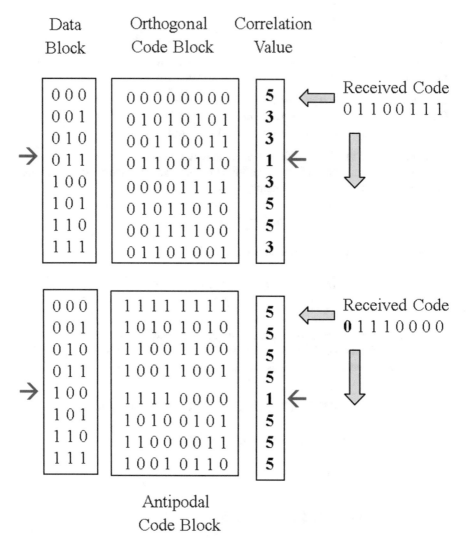

Fig. 3.14 ASK bandwidth at a glance. (**a**) Spectral response of NRZ data before modulation. (**b**) Spectral response of the carrier before modulation. (**c**) Spectral response of the carrier after modulation. The transmission bandwidth is $2f_b$, where f_b is the bit rate, and $T = 1/f_b$ is the bit duration for NRZ data

The spectral response is depicted in Fig. 3.15. Notice that the carrier frequency after FSK modulation varies back and forth from the nominal frequency f_c by $+ \Delta f_c$, where Δf_c is the frequency deviation. The FSK bandwidth is given by:

$BW = 2(f_b + \Delta f_c) = 2f_b(1 + \Delta f_c/f_b) = 2f_b(1 + \beta)$, where $\beta = \Delta f/f_b$ is known as the modulation index, and f_b is the coded bit frequency (bit rate R_b).

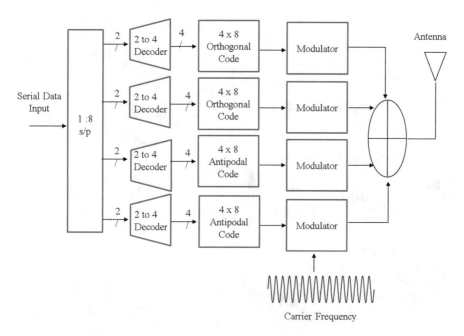

Fig. 3.15 FSK bandwidth at a glance. (**a**) Spectral response of NRZ data before modulation. (**b**) Spectral response of the carrier before modulation. (**c**) Spectral response of the carrier after modulation. The transmission bandwidth is $2(f_b + \Delta f_c)$. f_b is the bit rate, and Δf_c is the frequency deviation $= 1/f_b$ is the bit duration for NRZ data

3.7.5 BPSK Bandwidth at a Glance

In BPSK, the phase of the carrier changes in two discrete levels, in accordance to the input signal. Here we have:

Input data: $m(t) = 0$ or 1
Carrier frequency: $C(t) = A \cos(\omega t)$
Modulated carrier: $S(t) = A \cos(\omega + \varphi)t$

    where

- A = amplitude of the carrier frequency
- ω = angular frequency of the carrier
- $\varphi)$ = phase of the carrier frequency

The table below shows the number of phases and the corresponding bits per phase for MPSK modulation schemes for $M = 2, 4, 8, 16, 32, 64$, etc. It will be shown that higher-order MPSK modulation schemes ($M > 2$) are spectrally efficient. See Problem 3.7.

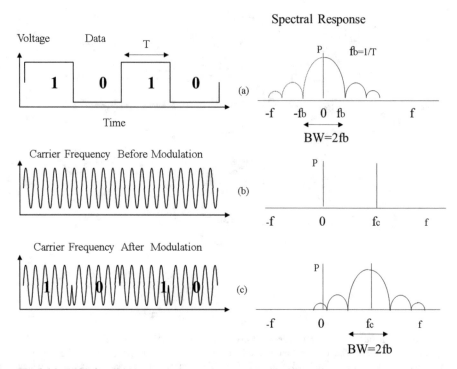

Fig. 3.16 BPSK bandwidth at a glance. (**a**) Spectral response of NRZ data before modulation. (**b**) Spectral response of the carrier before modulation. (**c**) Spectral response of the carrier after modulation

Modulation	Number of phases φ	Number of bits per phase
BPSK	2	1
QPSK	4	2
8PSK	8	3
16	16	4
32	32	5
64	64	6
:	:	:

Figure 3.16 shows the spectral response of the BPSK modulator. Since there are two phases, the carrier frequency changes in two discrete levels, one bit per phase, as follows:

$$\varphi) = 0^{\circ} \text{ for bit } 0$$

$$\varphi) = 180^{\circ} \text{ for bit } 1$$

Notice that the spectral response after BPSK modulation is the shifted version of the NRZ data, centered on the carrier frequency f_c. The transmission bandwidth is given by:

$$BW(BPSK) = 2R_b/\text{Bit per Phase} = 2R_b/1 = 2R_b$$

where

- R_b is the coded bit rate (bit frequency).
- For BPSK, $\varphi = 2$, one bit per phase.

Also, notice that the BPSK bandwidth is the same as the one in ASK modulation. This is due to the fact that the phase of the carrier changes in two discrete levels, while the frequency remains the same.

Problem 3.4: This Problem Relates to ASK

Given:

- Uncoded input bit rate: R_b (uncoded) $= 10$ kb/s
- Block coding: $M = 4, N = 4$
- Carrier frequency $f_c = 1$ MHz
- Modulation: ASK

Find:

(a) Code rate r
(b) Coded bit rate R_b(coded)
(c) Transmission bandwidth BW

Solution:

(a) $r = MN/(M + 1)(N + 1) = (4 \times 4)/(5 \times 5) = 16/25$
(b) Coded bit rate R_b (coded) $= R_b/r = 10$ kb/s$(25/16) = 15.625$ kb/s
(c) Transmission bandwidth:

$$BW = 2R_b \text{ (coded)} = 2 \times 15.625 \text{ kb/s} = 31.25 \text{ kb/s}$$

Problem 3.5: This Problem Relates to FSK
Given:

- Uncoded input bit rate: R_b (uncoded) $= 10$ kb/s
- Block coding: $M = 4, N = 4$
- Carrier frequency $f_c = 1$ MHz
- Modulation: FSK
- Modulation index $\beta = 1$

Find:

(a) Code rate r
(b) Coded bit rate R_b (coded)
(c) Transmission bandwidth BW

Solution:

(a) $r = MN/(M + 1)(N + 1) = (4 \times 4)/(5 \times 5) = 16/25$
(b) Coded bit rate R_b(coded) $= R_b/r = 10$ kb/s$(25/16) = 15.625$ kb/s
(c) Transmission bandwidth:

$$\text{BW} = 2R_b(1 + \beta) = 2 \times 15.625\text{kb/s}(1 + 1) = 62.5\text{kb/s}$$

Note: FSK needs more bandwidth.

Problem 3.6: This Problem Relates to BPSK
Given:

- Uncoded input bit rate: R_b(uncoded) $= 10$ kb/s
- Block coding: $M = 4, N = 4$
- Carrier frequency $f_c = 1$ MHz
- Modulation: BPSK (2 phases). 1 bit per phase

Find:

(a) Code rate r
(b) Coded bit rate R_b(coded)
(c) Transmission bandwidth BW

Solution:

(a) $r = MN/(M + 1)(N + 1) = (4 \times 4)/(5 \times 5) = 16/25$

(b) Coded bit rate $R_b(\text{coded}) = R_b/r = 10$ kb/s$(25/16) = 15.625$ kb/s

(c) Transmission bandwidth BW:

- Modulation is BPSK. Therefore, there are two phases, 1 bit per phase: $= 1$
- BPSK BW $= 2R_b(\text{coded})/\varphi = 2 \times 15.625$ kb/s$/1 = 31.25$ kb/s

Note: BPSK bandwidth is the same as in ASK.

Problem 3.7: This Problem Relates to QPSK ($Q = 4$)

Given:

- Uncoded input bit rate: $R_i(\text{uncoded}) = 10$ kb/s
- Block coding: $M = 4, N = 4$
- Carrier frequency $f_c = 1$ MHz
- Modulation: QPSK (4 phases), 2 bits per phase

Find:

(a) Code rate r

(b) Coded bit rate $R_b(\text{coded})$

(c) Transmission bandwidth BW

Solution:

(a) $r = MN/(M + 1)(N + 1) = (4 \times 4)/(5 \times 5) = 16/25$

(b) Coded bit rate $R_b = R_b/r = 10$ kb$/(25/16) = 15.625$ kb/s

(c) Transmission bandwidth:

$$BW = 2R_b(\text{coded})/\text{Bits Per Phase} = 2 \times 15.625(\text{kb/s})/2 = 15.625 \text{kHz}$$

Note:

- BPSk has 2 phases, 1 bit per phase: BW $= 2R_b/1 = 2R_b$ kHz
- QPSk has 4 phases, 2 bit per phase: BW $= 2R_b/2 = R_b$ kHz
- 8PSk has 8 phases, 3 bit per phase: BW $= 2R_b/3$ kHz
- 16PSk has 16 phases, 4 bit per phase: BW $= 2R_b/4 = R_b/2$ kHz
- And so on

Clearly, higher-order PSK modulation is bandwidth efficient.

Drill Exercise

Given:

- Uncoded input bit rate: R_b(uncoded) = 10 kb/s
- Block coding: $M = 6$, $N = 6$
- Carrier frequency $f_c = 1$ MHz
- Modulation: 64PSK

Find:

(a) Code rate r
(b) Coded bit rate R_b (coded)
(c) Transmission bandwidth BW

3.8   Conclusions

- The concept of block coding is presented in lucid language.
- Block code building blocks are presented to bring students up-to-date on key concepts and underlying principles in error control coding.
- Typical rectangular block coding is then presented with illustrations.
- Code rate and bandwidth are discussed with examples.
- A modified rectangular block coding is presented to improve error control capabilities.
- Modulation schemes are briefly presented to estimate the transmission bandwidth.
- Problems and exercises are inserted as needed.

References

1. G.C. Clark et al., *Error Correction Coding for Digital Communications* (Plenum press, New York, 1981)
2. G. Ungerboeck, Channel coding with multilevel/multiphase signals. IEEE Trans. Inf. Theory **IT28**, 55–67 (1982)
3. S. Lin, D.J. Costello Jr., *Error Control Coding: Fundamentals and Applications* (Prentice-Hall, Inc., Englewood Cliffs, 1983)
4. R.E. Blahut, *Theory and Practice of Error Control Codes* (Addison-Wesley, Reading, MA, 1983)
5. B. Sklar, *Digital Communications Fundamentals and Applications* (Prentice Hall, Upper Saddle River, 1988)

6. J.H. van Lint, *Introduction to Coding Theory*, GTM 86, 2nd edn. (Springer-Verlag, Berlin, 1992), p. 31. ISBN 3-540-54894-7
7. F.J. Mac Williams, N.J.A. Sloane, *The Theory of Error-Correcting Codes* (Elsevier, Amsterdam, 1977), p. 35. ISBN 0-444-85193-3
8. W. Huffman, V. Pless, *Fundamentals of Error-Correcting Codes* (Cambridge University Press, Cambridge, 2003) ISBN 978-0-521-78280-7
9. P.M. Ebert, S.Y. Tong, Convolutional Reed-Soloman codes. Bell Syst. Tech. J. **48**, 729–742 (1968)
10. R.G. Gallager, *Information Theory and Reliable Communications* (John Wiley, New York, 1968)
11. Kohlenbero, A. And Forney, G. D., Jr, Convolutional coding for channels with memory, IEEE Trans. Inf. Theory **IT-14**, 618–626. 1968
12. M.A. Reddy, S. M, Further results on convolutional codes derived from block codes. Inf. Control. **13**, 357–362 (1968)
13. S.M. Reddy, *A Class of Linear Convolutional Codes for Compound Channels*, Technical report (Bell Telephone Laboratories, Holmdel, 1968)
14. S.M. Reddy, J.P. And Robinson, A construction for convolutional codes using block codes. Inf. Control. **12**, 55–70 (1968)
15. M.J. Mueller, *A Bandwidth Efficient Modified Block Coded Modulation Technique*, MS Thesis, Department of Electrical Engineering, University of North Dakota, Grand Forks, North Dakota (2006)
16. D.R. Smith, *Digital Transmission Systems* (Van Nostrand Reinhold Co., 1985) ISBN: 0442009178
17. W. Leon, I.I. Couch, *Digital and Analog Communication Systems*, 7th edn. (Prentice-Hall, Inc., Englewood Cliffs, 2001) ISBN: 0-13-142492-0

Chapter 4
Convolutional Coding

Topics
- Introduction to Convolutional Coding
- Convolutional Code Building Blocks
- Construction of Convolutional Encoder
- Constraint Length, Code Rate and Bandwidth
- Construction of Convolutional Decoder
- Conclusions

4.1 Introduction

In convolutional coding, a sequence of data signals enters into the encoder, one bit at a time. The encoder generates n parity bits out of k information bits. The n parity bits, also known as coded information bits, are modulated and transmitted through a channel. At the receiver, the decoder recovers the data by means of code correlation. The ratio k/n is defined as the code rate r, where $r = k/n \leq 1$. The code rate is an indication of the amount of redundancy that protects the data. A low value for the code rate relates to more error-correcting ability but at the cost of increased bandwidth.

This type of error control is also classified as channel coding because these methods are often used to correct errors that are caused by channel noise. A typical rate ½ ($r = 1/2$) convolutional encoder is constructed as shown in Fig. 4.1, where,

- The information bits enters into the 3-bit shift register sequentially, one bit at a time.
- The convolutional encoder generates two parity bits for each entry of an information bit as encoded bits ($n > k$), where $r = k/n = 1/2$.
- The coded information bits are modulated and transmitted through a channel.

© The Editor(s) (if applicable) and The Author(s), under exclusive license to
Springer Nature Switzerland AG 2021
S. Faruque, *Free Space Laser Communication with Ambient Light Compensation*,
https://doi.org/10.1007/978-3-030-57484-0_4

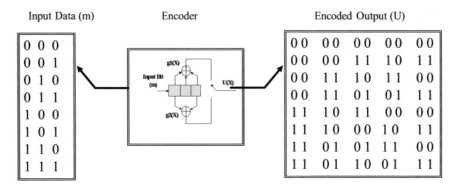

Input Data (m)   Encoder   Encoded Output (U)

Fig. 4.1 Illustration of a typical convolutional encoder. The information data serially enters into the 3-bit shift register, one bit at a time. The encoder generates 2 parity bits for each entry of an information bit. The code rate is defined as $r = k/n = 1/2$. The decoder recovers the data by means of code correlation

Table 4.1 Correlation receiver. Received data: 00 11 01 01 00

Input (m)	Output (U)		Correlation value
0. 0 0 0	00 00 00 00 00		4
1. 0 0 1	00 00 11 10 11		7
2. 0 1 0	00 11 10 11 00		3
3. 0 1 1	00 11 01 01 11		2
4. 1 0 0	11 10 11 00 00		5
5. 1 0 1	11 10 00 10 11		8
6. 1 1 0	11 01 01 11 00		4
7. 1 1 1	11 01 10 01 11		7

$m = 0\ 1\ 1$

Decoding is a process of code correlation as presented in Table 4.1. In this process, the receiver compares the received data with the expected data set to decode the actual data.

- A lookup table at the receiver contains the uncoded and the corresponding encoded data.
- Upon receiving an encoded data pattern, the receiver validates the received data pattern by means of code correlation.

This forms the basis of our presentation of FECC, based on convolutional coding [1–6]. In this chapter, we will present the key concepts, underlying principles, and practical application of convolutional coding schemes currently used in the telecommunication systems. Practical design and construction of convolutional coding schemes will be presented with illustrations. In particular, the following topics are presented in this chapter:

- Convolutional coding building blocks
- Typical convolutional coding
- Code rate and bandwidth
- Convolutional decoding

4.2   Convolutional Encoder Building Blocks

A convolutional encoder, in its most basic construction consists, of three building blocks as listed below:

- Shift Register
- Exclusive OR Gates
- Multiplexer

A brief description of each of these building blocks are presented below.

4.2.1   Shift Register (SR)

A shift register (SR) is a device that converts serial data into parallel formats or vice versa. In the case of serial to parallel SR, data enters into the SR serially, one bit at a time. Once the data has been clocked in, the content of the SR can be read off at each output simultaneously for further processing. Fig. 4.2 illustrates the operation of a 3-bit serial to parallel shift register, which will be used to construct a convolutional encoder.

Let's assume that the input bit sequence $m = 1\ 0\ 1$, where the first entry is 1, second entry is 0, and the third entry is 1: also, we assume that the initial content of the SR is 0 0 0.

At $t = 0$ (Fig. 4.2a):

- Initial content of the SR is 0 0 0
- Parallel output is 0 0 0

At $t = 1$(Fig. 4.2b):

- The first bit entry into SR $= 1$
- Content of SR: 1 0 0
- Parallel output is 1 0 0

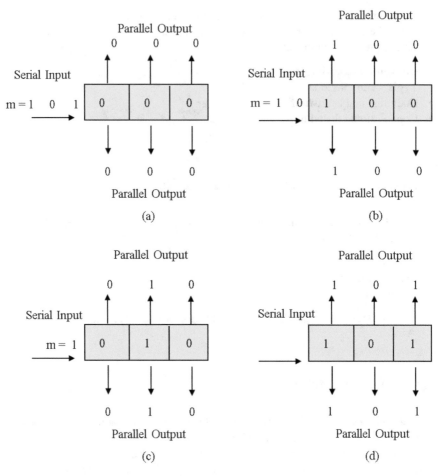

Fig. 4.2   Operation of a 3-bit serial to parallel shift register

At $t = 2$ (Fig. 4.2c):

- The second bit entry into SR $= 0$
- The first bit moves forward by one bit and occupies the next location of the SR
- The content of SR: 0 1 0
- Parallel output is 0 1 0

At $t = 3$ Fig. 4.2d):

- The third bit entry into SR $= 1$
- The previous two bits move forward by one bit
- Content of SR is 1 0 1
- Parallel output is 1 0 1

4.2.2 Exclusive OR Gates as Parity Generators

A parity generator is an array of Exclusive OR (EXOR) gates that generate parity bits known as odd parity or even parity. These parity generators are used to construct the convolutional encoder in conjunction with a shift register. Our objective now is to examine exclusive OR gates and observe how parity bits can be generated by inspection. This is governed by the following logic:

INPUT = EVEN 1's $\rightarrow$ OUTPUT = 0
INPUT = ODD 1's $\rightarrow$ OUTPUT = 1

This analogy is used to derive parity values by inspection as shown in Fig. 4.3.

4.2.3 Symbolic Representation of Exclusive OR Gates

Figures 4.4 and 4.5 show symbolic representation and examples of Exclusive OR Gates, where the parity estimations are based on inspection. Once again, this is governed by the following logic

INPUT = EVEN 1's $\rightarrow$ OUTPUT = 0
INPUT = ODD 1's $\rightarrow$ OUTPUT = 1

Fig. 4.3 Generation of parity bits by inspection

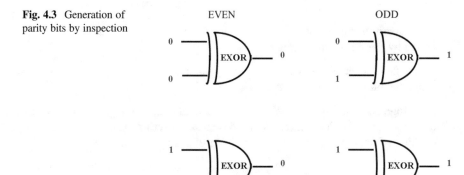

Fig. 4.4 Symbolic representation of exclusive OR gates

Fig. 4.5 Examples of parity
estimates by inspection

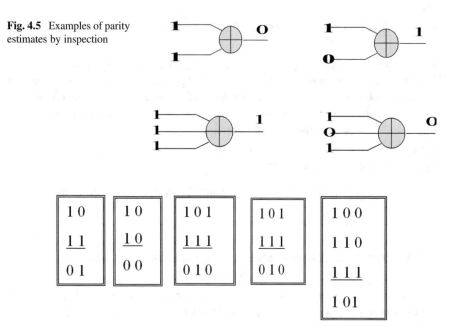

Fig. 4.6 Examples of MOD2 operation. This is obtained by counting the number of 1's vertically

4.2.4 Modulo-2 Addition (MOD-2 ADD)

Parity is an arithmetic operation, also known as Modulo2 or MOD2 addition. This is
governed by the following analogy:

- When the number of 1's is even, the parity value is 0.
- When the number 1's is odd, the parity value is 1.

Therefore, by counting the number of 1's, the parity value of a given word can be
determined simply by inspection. Figure 4.6 shows a set of examples to illustrate
this, which is obtained by counting the number of 1's vertically.

4.2.5 Multiplexers

Multiplexing, also commonly referred to as MUX, is a method of transmitting and
receiving multiple independent signals over a single transmission channel serially in
a preassigned time slot. MUX at the transmit side assigns multiple channels in
pre-assigned time slots. MUX at the receive side, known as the de-multiplexer

Fig. 4.7 Symbolic representation of a 2:1 multiplexer

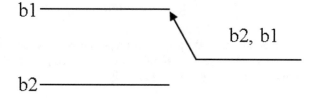

(DEMUX), separates the incoming composite signal into parallel streams. Both multiplexer and de-multiplexer are synchronized by a common clock to receive data in accordance with the transmit sequence.

Figure 4.7 shows a symbolic representation of a 2:1 multiplexer, which will be used in the construction of a convolutional encoder. Here, 2:1 represents two input bit streams converted into a single bit stream.

Problem 4.1

Given:

Two bit sequences $b1$ and $b2$ are as follows:

$b1 = 1\ 0\ 1$
$b2 = 1\ 1\ 1$

(a) Find the output bit sequences as $b1, b2, \ldots$
(b) If the input bit rate is 10 kb/s, calculate output bit rate.

Solution:

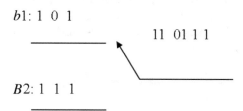

Therefore, the output bit sequence is: 11 01, 11

(b) Output bit rate = input rate × 2 = 10 × 2 = 20 kb/s

4.2.6   *Polynomial Representation of Data*

In convolutional coding, we often represent data by means of polynomials and their products [7]. Here is a simple example. A sequence of bits $m = 1\ 0\ 1$ can be expressed as polynomial $m(x)$ as follows:

$$m(X) = 1 + 0.X + 1.X^2$$
$$= 1 + X^2$$

Similarly, another sequence of bits $g = 1\ 1\ 1$ can be expressed as polynomial $g(x)$ as follows:

$$g(X) = 1 + 1.X + 1.X^2$$
$$= 1 + X + X^2$$

Then the product of the above two polynomials can be written as:

$$m(X)g(X) = \left(1 + X^2\right)\left(1 + X + X^2\right)$$
$$= 1 + X + X^2 + X^2 + X^3 + X^4$$
$$= 1 + X + X^3 + X^4$$
$$= 1 + 1.X + 1.X2 + 1.X3 + 1.X4$$
$$= 1\ 1\ 1\ 1\ 1$$

where $X^2 + X^2 = 0$ (MOD-2 Addition).

This forms the basis of representing data by means of polynomials and the product of two polynomials. In this chapter, we will further examine this method and show how convolutional encoders can be designed, constructed, and verified by means of inspection.

Problem 4.2

This problem illustrates how to represent data by means of polynomials and their products.

Given:

$$m = 110$$
$$g = 1\ 0\ 0$$

Find:

(a) $m(x)$
(b) $g(x)$
(c) $m(x)g(x)$

Solution:

(a) $m = 110$: $m(x) = 1 + .X + 0.X^2 = 1 + X2$
(b) $g = 1\ 0\ 0$: $g(x) = 1 + 0.X + 0.X2 = 1$
(c) $m(x)g(x) = (1 + X2)0.1 = 1 + x2 = 1 + 0.X + 1.X2 = 1\ 0\ 1$

4.3   Construction and Operation of Convolutional Encoder

A convolutional encoder, in its most basic construction, is shown in Fig. 4.8. It contains a shift register (SR), an exclusive OR gate known as the upper generator ($g1$), a second exclusive OR gate known as lower generator ($g2$) and a 2:1 multiplexer. Here, a sequence of data bits enters into the 3-bit SR one bit at a time, and a corresponding 3-bit parallel data is used to generate two parity bits through a pair of parity generators $g1$ and $g2$. The encoded output data is then taken serially by means of a 2:1 multiplexer as $u1$ and $u2$.

Briefly, the operation of the encoder is as follows:

- The initial content of the 3-bit shift register (SR) $= 0\ 0\ 0$.
- The input bit enters into the SR serially, one bit at a time.

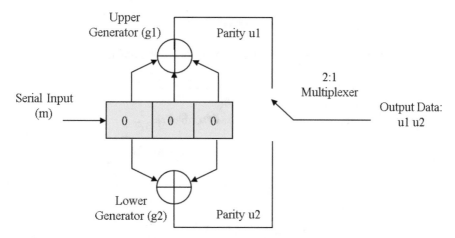

Fig. 4.8 A convolutional encoder containing a shift register (SR), an exclusive OR gate known as the upper generator ($g1$), a second exclusive OR gate known as lower generator ($g2$) and a 2:1 multiplexer

- For each single bit entry into the SR, the content of the SR is updated.
- The upper generator $g1$ generates a parity bit $u1$, and the lower generator $g2$ generates a parity bit $u2$.
- The output is 2:1 time division multiplexed to obtain: $u1$ and $u2$.
- Next, a new bit enters into the SR, and a new pair of $u1u2$ is generated at the output.
- The process continues. Consequently, for every single bit entry into the SR, there are two output parity bits, resulting in a rate 1/2 ($r = 1/2$) convolutional encoder.

4.3.1 Polynomial Method of Analysis

Let's consider Fig. 4.8 again and assume that the input bit sequence is 1 0 1, where the first entry is 1, the second entry is 0, and the third entry is 1. Also, we assume that the initial content of the SR is 0 0 0. The encoder is described to have:

- Input data: $m = 1\,0\,1$
- Upper generator $g1 = 1\,1\,1$ (according to the input connectivity)
- Lower generator $g2 = 1\,0\,1$ (according to the input connectivity)

The input sequence $m = 1\,0\,1$ is described by the following polynomial:

$$m(X) = 1 + 0X + 1X^2$$
$$= 1 + X^2$$

The upper generator is described by the following polynomial:

$$g1(X) = 1 + X + X^2$$

The lower generator is described by the following polynomial:

$$g2(X) = 1 + X^2$$

Then the product of polynomials can be described as:

$$m(X)g1(X) = \left(1 + X^2\right)\left(1 + X + X^2\right)$$
$$= 1 + X + X^3 + X^4$$

$$m(X)g2(X) = \left(1 + X^2\right)\left(1 + X^2\right)$$
$$= 1 + X^4$$

With $X^2 + X^2 = 0$, the output bit sequence can be found as $U(X) = m(X)g1(X)$ multiplexed with $m(X)g2(X)$, where $m(X)$ is the input bit sequence [7]. We write the above two equations as follows:

$$m(X)g1(X) = 1 \ + \ 1X \ + \ 0X^2 \ + \ 1X^3 \ + \ 1X^4$$

$$m(X)g2(X) = 1 \ + \ 0X \ + \ 0X^2 \ + \ 0X^3 \ + \ 1X^4$$

--- ---------------

$$U(X) = (1,1) + (1,0) \ + \ (0,0) + \ (1,0) \ + (1,1)$$

Taking only the coefficients, we obtain the desired multiplexed output bit sequence as follows:

$$U = 11 \ 10 \ 00 \ 10 \ 11$$

4.3.2   Verification by Inspection

Let's consider Fig. 4.9 and observe what happens when a single bit enters into the SR and moves through the SR, one bit at a time: Let's also assume that the input bit sequence is 1 0 1, where the first entry is 1, the second entry is 0, and the third entry is 1. Also, we assume that the initial content of the SR is 0 0 0. Here, we have:

- Input data: $m = 1 \ 0 \ 1$
- Generator $g1 = 1 \ 1 \ 1$
- Generator $g2 = 1 \ 0 \ 1$

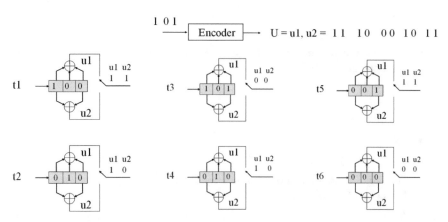

Fig. 4.9 Step-by-step operation of convolutional encoder. For each entry of a bit, the output generates two parity bits

Now, observe the output as the data enters into the encoder, one bit at a time.

At $t = 0$:

- Initial content of the SR is 0 0 0
- $m(x) = 0$
- $g(x) = 0$
- The output is: u1 u2 $= 0\ 0$

At $t = 1$:

- The first bit entry into SR $= 1$
- Content of SR: 1 0 0
- $m(x) = 1$
- $g(x) = 1$
- U1u2 $= 1\ 1$

At $t = 2$:

- The second bit entry into SR $= 0$
- The first bit moves forward by one bit and occupies the next location of the SR
- The content of SR: 0 1 0
- $m(x) = 1$
- $g(x) = 0$
- U1u2 $= 1\ 0$

At $t = 3$:

- The third bit entry into SR $= 1$
- The previous two bits move forward by one bit
- The content of SR: 1 0 1
- $m(x) = 0$
- $g(x) = 0$
- U1u2 $= 0\ 0$

At $t = 4$:

- The fourth bit entry into SR $= 0$
- The previous two bits move forward by one bit
- The content of SR: 0 1 0
- $m(x) = 1$
- $g(x) = 0$
- U1u2 $= 1\ 0$

At $t = 5$:

- The fifth bit entry into SR $= 0$
- The previous two bits move forward by one bit
- The content of SR: 0 0 1
- $m(x) = 1$
- $g(x) = 1$
- U1u2 $= 1\ 1$

At $t = 6$:

- The sixth bit entry into SR $= 0$
- The previous two bits move forward by one bit
- The SR is now cleared (three bits are out)
- The content of SR: 0 0 0 (SR is initialized)
- $m(x) = 0$
- $g(x) = 0$
- U1u2 $= 0\ 0$ (back to initial condition, therefore ignore this pair of data)

The operation is now complete. The encoded output bit sequence is:

$$U = 11\ 10\ 00\ 10\ 11\ 00$$

Note that the last two bits are 0 0, which is the initial condition of the SR. Therefore, these two bits are not considered and neglected. Also notice that the outcome is the same as those obtained earlier by means of the polynomial method.

4.3.3 Constraint Length, Code Rate, and Bandwidth

The encoder we have just described is said to have the following parameters:

- Constraint length $k = 3$ (This is the length of the SR)
- Code Rate $r = 1/2$

Typical Parameters:

- $k = 7$ and 9
- $r = 1/2,\ 1/3,\ 2//3,\ 3/4$, etc.

The code rate (r) is defined as:

- $r = k/n\ (r < 1)$
- $k =$ Number of information bits
- $n =$ Number of encoded information bits

The bandwidth is defined as:

- BW - $= R_b/r$

where

- $R_b =$ uncoded bit rate b/s
- $R =$ code rate

Problem 4.3

This problem shows how to analyze a given convolutional encoder by polynomial method and verify by means of inspection.

Given:

- The input bit sequence: $m = 0\ 0\ 1$
- Constraint length $k = 3$
- Upper generator $g1 = 1\ 1\ 1$
- Lower Generator $g2 = 1\ 0\ 1$
- Constraint length $k = 3$

Find:

(a) The output bit sequence U
(b) What is the code rate:
(c) If the input bit rate is 10 kb/s, what is the encoded output bit rate?

Solution:

$$m(X) = X^2$$

$$g1(X) = 1 + X + X^2$$

The product of these polynomials is given by:

$$m(X)g1(X) = (X^2)(1 + X + X^2)$$
$$= X^2 + X^3 + X^4$$

Similarly, we obtain the product of two polynomials as:

$$m(X)g2(X) = (X^2)(1 + X^2)$$
$$= X^2 + X^4$$

The above two product of polynomials can be written as follows:

$$m(X)g1(X) = 0 + 0X + 1X^2 + 1X^3 + 1X^4$$
$$m(X)g2(X) = 0 + 0X + 1X^2 + 0X^3 + 1X^4$$

$$U(X) = (0,0) + (0,0) + (1,1) + (1,0) + (1,1)$$

The encoded output bit sequence is then obtained by taking the coefficients:

$$U = 00\ 00\ 11\ 10\ 11$$

Verification by inspection [8]:

$m =$ 0   0   1
         ($t1$)   ($t2$)   ($t3$)
1st entry $= 0$ (at $t1$)
2nd entry $= 0$ (at $t2$)
3rd entry $= 1$ (at $t3$)
Remaining entries are all zeros.

The encoded output bit sequence U can be obtained simply by inspection as given below:

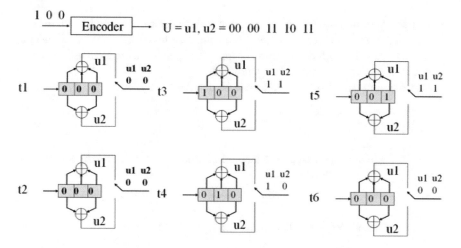

It may be noted that, the last two bits are 0 0, which is the initial condition of the SR. Therefore, these two bits are not considered and neglected. That the outcome is the same as those obtained earlier by means of polynomial method.

Problem 4.4
This problem relates to code rate and bandwidth.

Given:

- A rate ½ convolutional encoder
- Input bit rate $R_b = 10$ kb/s
- Non-return to zero (NRZ) data

Find:

(a) The encoded bit rate
(b) Bandwidth (BW) associated with the encoded data

Solution:

(a) Encoded bit Rate $= R_b/r = 2R_b = 2 \times 10\text{kb/s} = 20$ kb/s
(b) BW $= 2 \times$ Encoded Bit Rate $= 2 \times 20\text{kb/s} = 40$ kHz

4.4 Summary of Convolutional Encoder

A rate $r = 1/2$, constraint length $k = 3$ Convolutional encoder is described by the structure as shown in Fig. 4.10. It contains a 3-bit shift register, an upper parity generator $g1$, a lower parity generator $g2$, and a 2:1 multiplexer. The operation is as follows:

- A 3-bit input bit sequence enters into the shift register (SR), one bit at a time.
- For each 3-bit input bit pattern, there is a unique 10-bit encoded output bit pattern.
- Since there are $2^3 = 8$ combination of input data pattern, there are 8 unique 10-bit encoded bit patterns available at the output.
- These output encoded bit patterns can be determined by:

  – Polynomial method
  – By inspection

In the following, the inspection method is used to determine these encoded bit patterns.

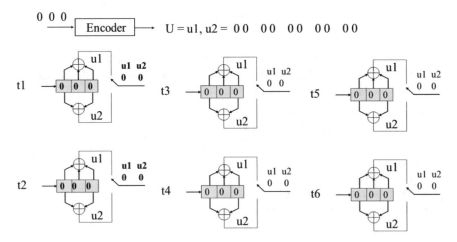

Fig. 4.10 Generation of encoded output bit pattern U for an input bit pattern $m = 0\ 0\ 0$

Input (*m*): 0 0 0:

Input (*m*): 0 0 1 (Fig. 4.11)

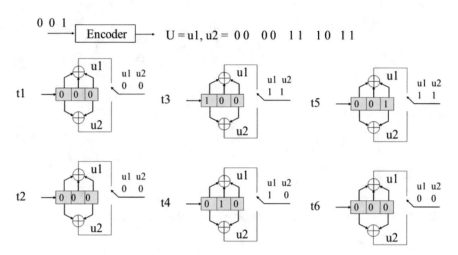

Fig. 4.11   Generation of encoded output bit pattern *U* for an input bit pattern *m* = 0 0 1

Input (*m*): 0 1 0 (Fig. 4.12)

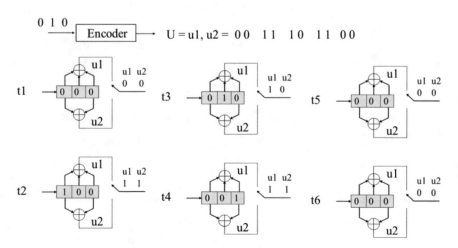

Fig. 4.12   Generation of encoded output bit pattern *U* for an input bit pattern *m* = 0 1 0

Input (_m_): 0 1 1 (Fig. 4.13)

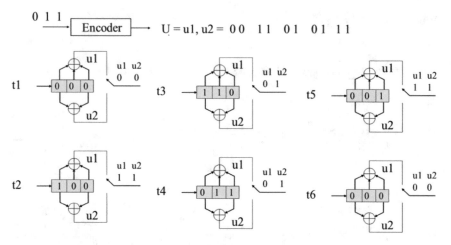

Fig. 4.13   Generation of encoded output bit pattern U for an input bit pattern $m = 0\ 1\ 1$

Input (_m_): 1 0 0 (Fig. 4.14)

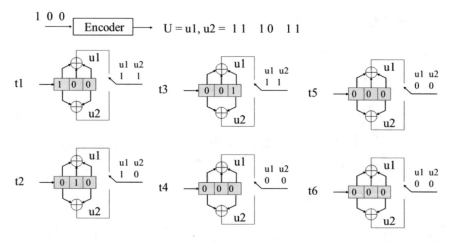

Fig. 4.14   Generation of encoded output bit pattern U for an input bit pattern $m = 1\ 0\ 0$

Input (*m*): 1 0 1 (Fig. 4.15)

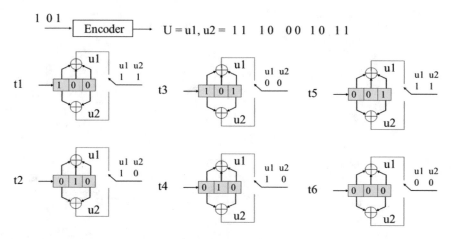

Fig. 4.15   Generation of encoded output bit pattern *U* for an input bit pattern *m* = 1 0 1

Input (*m*): 1 1 0 (Fig. 4.16)

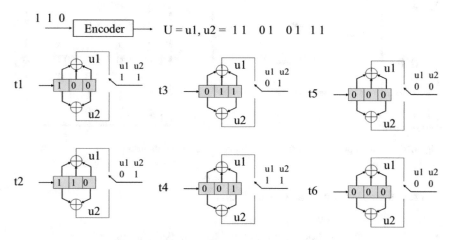

Fig. 4.16   Generation of encoded output bit pattern *U* for an input bit pattern *m* = 1 1 0

Input (*m*): 1 1 1 (Fig. 4.17)

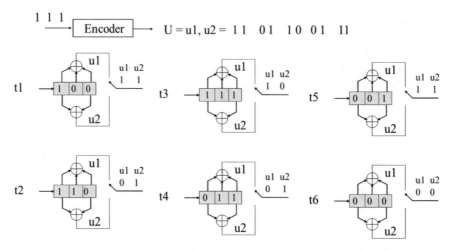

Fig. 4.17 Generation of encoded output bit pattern U for an input bit pattern $m = 1\ 1\ 1$

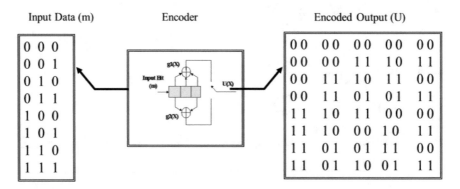

Fig. 4.18 Input/output mapping for $k = 3$ and $r\text{-} = 1/2$ convolutional encoder

From the above illustrations, we see that

- 3-bit data has $2^3 = 8$ combinations.
- Each input combination generates a unique encoded bit pattern.
- These encoded bits are modulated and transmitted through a channel.

Figure 4.18 shows the encoder input/output mapping for $k = 3$, $r = 1/2$.

Table 4.2 Lookup table a

Input (m)	Output (U)
0. 0 0 0	00 00 00 00 00
1. 0 0 1	00 00 11 10 11
2. 0 1 0	00 11 10 11 00
3. 0 1 1	00 11 01 01 11
4. 1 0 0	11 10 11 00 00
5. 1 0 1	11 10 00 10 11
6. 1 1 0	11 01 01 11 00
7. 1 1 1	11 01 10 01 11

4.5 Convolutional Decoder

4.5.1 Generation of a Lookup Table

Decoding is a process of code correlation. In this process, the receiver compares the received data with the expected data set to recover the actual data. The expected data is stored into a lookup table. (Table 4.2) [8]:

- The lookup table at the receiver contains the input/output bit sequences.
- For $m = 3$, there are 8 possible output combinations of 3-bit data.
- For each combination of 3-bit data, there is a unique encoded 10-bit data (see table).
- The receiver receives one of 8 output sequences.
- Upon receiving an encoded data pattern, the receiver validates the received data pattern by means of code correlation.

The correlation process and validation of the received data are presented in the following section.

4.5.2 Code Correlation Process

Let's examine the correlation process using the following example:

The input bit pattern $m = 0\ 1\ 1$
Encoded transmit data: $U = 00\ 11\ 01\ 01\ 11$
Received data with errors $U^* = 00\ 11\ 01\ 01\ \mathbf{0\ 0}$

Notice that the last two bits are in error, identified in bold. Now, let's determine how the receiver recovers the correct data, where the actual input data is $m = 0$ 1 1. This is a correlation process, requiring several tests to validate the actual data. The correlation process is described below:

Test 0

This test compares the received data with the first row of data stored in the lookup table and counts the number of positions it does not match. This is accomplished by MOD2 operation (EXOR operation). The result is presented below:

Received Data: 00 11 01 01 **00**
1st row of data in the lookup table: 00 00 00 00 00
Mod-2 Add   00 11 01 01 00
Correlation Value $= 4$ (count the number of 1's in MOD2 Add)
Verdict: No match, Continue search.

Input (m)	Output (U)	Correlation Value		Received Data
0. 0 0 0	00 00 00 00 00	4	←	00 11 01 01 00
1. 0 0 1	00 00 11 10 11			
2. 0 1 0	00 11 10 11 00			
3. 0 1 1	00 11 01 01 11			
4. 1 0 0	11 10 11 00 00			
5. 1 0 1	11 10 00 10 11			
6. 1 1 0	11 01 01 11 00			
7. 1 1 1	11 01 10 01 11			

Test 1

This test compares the received data with the second row of data stored in the lookup table and counts the number of positions it does not match. This is accomplished by MOD2 operation (EXOR operation). The result is presented below:

Received Data: 0 0 11 01 01 00
2nd row of data in the lookup table: 0 0 00 11 10 11
Mod-2 Add: 0 0 1 1 10 111 11
Correlation value $= 7$
Verdict: No match, Continue search

Input (m)	Output (U)	Correlation Value
0. 0 0 0	00 00 00 00 00	
1. 0 0 1	00 00 11 10 11	7
2. 0 1 0	00 11 10 11 00	
3. 0 1 1	00 11 01 01 11	
4. 1 0 0	11 10 11 00 00	
5. 1 0 1	11 10 00 10 11	
6. 1 1 0	11 01 01 11 00	
7. 1 1 1	11 01 10 01 11	

Received Data

00 11 01 01 00

Test 2

This test compares the received data with the third row of data stored in the lookup table and counts the number of positions it does not match. This is accomplished by MOD2 operation (EXOR operation). The result is presented below:

Received Data: 00 11 01 01 00
3rd row of data in the lookup table: 00 11 10 11 00
Mod-2 Add: 00 00 11 10 00
Correlation Value $= 3$
Verdict: No match, Continue search

Input (m)	Output (U)	Correlation Value
0. 0 0 0	00 00 00 00 00	
1. 0 0 1	00 00 11 10 11	
2. 0 1 0	00 11 10 11 00	3
3. 0 1 1	00 11 01 01 11	
4. 1 0 0	11 10 11 00 00	
5. 1 0 1	11 10 00 10 11	
6. 1 1 0	11 01 01 11 00	
7. 1 1 1	11 01 10 01 11	

Received Data

00 11 01 01 00

Test 3

This test compares the received data with the fourth row of data stored in the lookup table and counts the number of positions it does not match. This is accomplished by MOD2 operation (EXOR operation). The result is presented below:

Received Data: 0 0 11 01 01 00
4th row in the lookup table: 0 0 11 01 01 11
Mod-2 Add: 0 0 0 0 00 00 11
Correlation Value: 2
Verdict: Possible Candidate

Input (*m*)	Output (*U*)	Correlation Value
0. 0 0 0	00 00 00 00 00	
1. 0 0 1	00 00 11 10 11	
2. 0 1 0	00 11 10 11 00	
3. 0 1 1	00 11 01 01 11	2
4. 1 0 0	11 10 11 00 00	
5. 1 0 1	11 10 00 10 11	
6. 1 1 0	11 01 01 11 00	
7. 1 1 1	11 01 10 01 11	

Received Data

00 11 01 01 00

Test 4

This test compares the received data with the fifth row of data stored in the lookup table and counts the number of positions it does not match. This is accomplished by MOD2 operation (EXOR operation). The result is presented below:

Received Data: 00 11 01 01 00
5TH row in the lookup table: 11 10 11 00 00
Mod-2 Add: 11 01 10 01 00
Correlation Value = 5
Verdict: No match, Continue Search

Input (m)	Output (U)	Correlation Value
0. 0 0 0	00 00 00 00 00	
1. 0 0 1	00 00 11 10 11	
2. 0 1 0	00 11 10 11 00	
3. 0 1 1	00 11 01 01 11	
4. 1 0 0	11 10 11 00 00	5
5. 1 0 1	11 10 00 10 11	
6. 1 1 0	11 01 01 11 00	
7. 1 1 1	11 01 10 01 11	

Received Data

00 11 01 01 00

Test 5

This test compares the received data with the sixth row of data stored in the lookup table and counts the number of positions it does not match. This is accomplished by MOD2 operation (EXOR operation). The result is presented below:

Received Data: 00 11 01 01 00
6th row in the lookup table: 11 10 00 10 11
Mod-2 Add: 11 01 01 11 11
Correlation Value = 8
Verdict: No match, Continue Search

Input (m)	Output (U)	Correlation Value
0. 0 0 0	00 00 00 00 00	
1. 0 0 1	00 00 11 10 11	
2. 0 1 0	00 11 10 11 00	
3. 0 1 1	00 11 01 01 11	
4. 1 0 0	11 10 11 00 00	
5. 1 0 1	11 10 00 10 11	8
6. 1 1 0	11 01 01 11 00	
7. 1 1 1	11 01 10 01 11	

Received Data

00 11 01 01 00

Test 6

This test compares the received data with the seventh row of data stored in the lookup table and counts the number of positions it does not match. This is accomplished by MOD2 operation (EXOR operation). The result is presented below:

Received Data: 00 11 01 01 00
7th row in the lookup table: 11 01 01 11 00
Mod-2 Add: 11 10 00 10 00
Count the no, of 1's = 4
Verdict: No match. Continue search

Input (*m*)	Output (*U*)	Correlation Value
0. 0 0 0	00 00 00 00 00	
1. 0 0 1	00 00 11 10 11	
2. 0 1 0	00 11 10 11 00	
3. 0 1 1	00 11 01 01 11	
4. 1 0 0	11 10 11 00 00	
5. 1 0 1	11 10 00 10 11	
6. 1 1 0	11 01 01 11 00	4
7. 1 1 1	11 01 10 01 11	

Received Data

00 11 01 01 00

Test 7

This test compares the received data with the eighth row of data stored in the lookup table and counts the number of positions it does not match. This is accomplished by MOD2 operation (EXOR operation). The result is presented below:

Received Data: 00 11 01 01 00
7th row in the lookup table: 11 01 10 01 11
Mod-2 Add: 11 10 11 00 11
Correlation Value = 7
Verdict: No Match
Test is complete

Input (*m*)	Output (*U*)	Correlation Value
0. 0 0 0	00 00 00 00 00	
1. 0 0 1	00 00 11 10 11	
2. 0 1 0	00 11 10 11 00	
3. 0 1 1	00 11 01 01 11	
4. 1 0 0	11 10 11 00 00	
5. 1 0 1	11 10 00 10 11	
6. 1 1 0	11 01 01 11 00	
7. 1 1 1	11 01 10 01 11	7

Received Data

00 11 01 01 00

The Final Verdict

Collect the correlation values and validate the data that indicates the lowest correlation value. This is presented in the following lookup table.

Final Verdict: Look Up Table

Input (*m*)	Output (*U*)	Correlation Value
0. 0 0 0	00 00 00 00 00	4
1. 0 0 1	00 00 11 10 11	7
2. 0 1 0	00 11 10 11 00	3
3. 0 1 1	00 11 01 01 11	2
4. 1 0 0	11 10 11 00 00	5
5. 1 0 1	11 10 00 10 11	8
6. 1 1 0	11 01 01 11 00	4
7. 1 1 1	11 01 10 01 11	7

Accept ⟶ (row 3)   Accept ⟵

In examining the above table, we find that:

- The lowest correlation value is 2.
- The corresponding data is $m = 0\ 1\ 1$.
- This is the data which has been transmitted to the receiver.

4.5.3   A Further Note on Code Rate

- Each 3-bit input data is preceded by three zero bits, for a total of 6 input bits.
- The total number of encoded bits is 12, where the last two encoded bits are 0 0, which have been neglected.
- Therefore, the code rate is: $r = 6/12 = 1/2$.

Viewed from another angle, we observe that for each entry of an input bit, the encoder generates two parity bits at the output, indicating that the code rate is also ½.

4.6   Conclusions

In this chapter, we have presented the key concepts, underlying principles, and practical application of convolutional coding schemes currently used in the telecommunication systems. Practical design and construction of convolutional coding along with decoding schemes are presented with illustrations. In particular, the following topics are presented in this chapter:

- Convolutional coding building blocks
- Typical convolutional coding
- Convolutional decoding
- Code rate and bandwidth

References

1. P.M. Ebert, S.Y. Tong, Convolutional Reed-Soloman codes. Bell Syst. Tech. J. **48**, 729–742 (1968)
2. R.G. Gallager, *Information Theory and Reliable Communications* (John Wiley, New York, 1968)
3. Kohlenbero, A. And Forney, G. D., Jr, "Convolutional coding for channels with memory", IEEE Trans. Inf. Theory **IT-14**, 618–626. 1968
4. S.M. Reddy, Further results on convolutional codes derived from block codes. Inf. Control. **13**, 357–362 (1968)

5. S.M. Reddy, *A Class of Linear Convolutional Codes for Compound Channels*, Technical Report (Bell Telephone Laboratories, Holmdel, 1968)
6. S.M. Reddy, J.P. And Robinson, A construction for convolutional codes using block codes. Inf. Control. **12**, 55–70 (1968)
7. W. Leon, I.I. Couch, *Digital and Analog Communication Systems*, 7th edn. (Prentice-Hall, Inc., Englewood Cliffs, 2001) ISBN: 0-13-142492-0
8. S. Faruque, *Lecture Notes, EE411: Communications Engineering* (EE Department, University of North Dakota, Grand Forks, 2013)

Chapter 5
Waveform Coding

Topics

- Introduction to Waveform Coding
- Conceptual Development
- Orthogonal and Antipodal Codes
- Error Control Coding Based on Orthogonal Codes
- Waveform Coding and Decoding
- Waveform Capacity
- Conclusions

5.1 Introduction

Waveform coding is a form of channel coding where a set of waveforms is transformed into a set of orthogonal waveforms, so that the detection process has fewer errors [1]. There are two classes of waveform coding: (a) M-ary signaling and (b) orthogonal coding.

In M-ary signaling, a k-bit data set, is used to address $M=2^k$ modulated waveforms (e.g., MFSK). This process provides improved error performance at the expense of bandwidth. Figure 5.1 illustrates a typical M-ary signaling scheme, requiring $M=16$ waveforms, to transmit $k=4$ bit data. It is bandwidth inefficient.

Similarly, in bi-orthogonal coding, a k-bit data set, is directly mapped into $2n$ bi-orthogonal codes, where n is the code length. This is shown in Fig. 5.2, where a block of 4-bit data is mapped into a block of 16-bi-orthogonal data set. This approach is also bandwidth inefficient, since the k-bit data set is directly mapped into 2×2^k bi-orthogonal codes.

In this chapter, we present an alternate method of waveform coding based on orthogonal codes, which does not consume additional bandwidth and offers protection against errors [2–4]. In the proposed method, conceptually shown in Fig. 5.3, a

Data

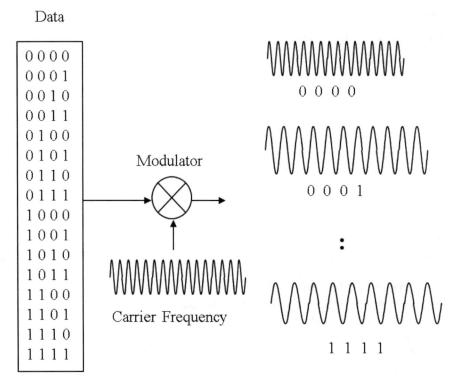

Fig. 5.1 Typical *M*-ary signaling, requiring 16 waveforms to transmit 4-bit data. It is bandwidth inefficient

high-speed data stream is inverse multiplexed into several parallel streams. These parallel streams, now reduced in speed, are grouped into a number of subsets and mapped into a predetermined group of bi-orthogonal codes and then modulated by a bank of modulators using the same carrier frequency. This methodology substantially reduces the required number of waveforms and enhances transmission efficiency.

Our objective is to show that orthogonal codes are essentially (n, k) block codes where a k-bit information is represented by a unique n-bit orthogonal code ($k < n$). We examine this by noting that an n-bit orthogonal code has $n/2$ 1s and $n/2$ 0s, i.e., there are $n/2$ positions where 1s and 0s differ. Therefore, the distance between two orthogonal codes is also $n/2$. This distance property can be exploited to achieve bandwidth efficient forward error control coding (FECC). We show that an n-bit orthogonal code can correct t errors where $t = (n/4)-1$, where n is the code length. A measure of coding gain is then obtained by comparing the word error with coding to the word error without coding.

The bandwidth efficiency is achieved by inverse multiplexing the base band binary data into several parallel streams. These parallel streams, now reduced in

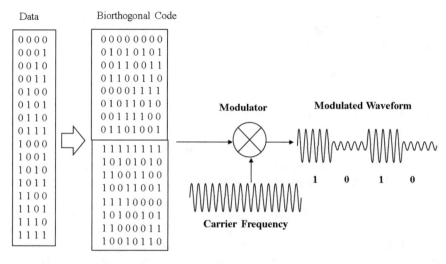

Fig. 5.2 Typical bi-orthogonal coding, requiring longer code to represent a 4-bit data. Which is also bandwidth inefficient

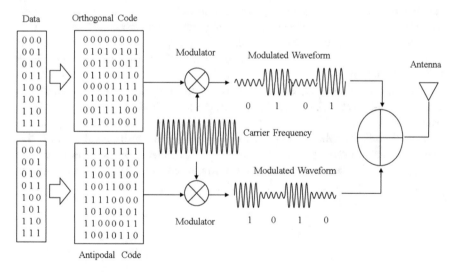

Fig. 5.3 Proposed waveform coding, which is bandwidth efficient

speed, are partitioned into a number of data blocks. Each subset of data is then used to address a predetermined subset of bi-orthogonal codes, stored in a ROM (Read Only Memory). A bank of identical modulators subsequently modulates the corresponding code, combines, and transmits through a given channel. This methodology achieves a code rate $r = k/n$ where n is the code length, and k is the data set. It follows that code rates such as rate 1/2, rate 3/4, rate 1, etc. are indeed available out of orthogonal codes with bandwidth efficiency. Construction of rate ½, rate ¾, and rate 1 orthogonal coded modulation schemes, using 8-bit and 16-bit bi-orthogonal code, is presented to illustrate the concept

5.2   Conceptual Development

Figure 5.4 briefly illustrates the concept. In Fig. 5.4a we have the conventional method of waveform coding where a 4-bit data set is represented by $2^4=16$ waveforms. This scheme is commonly viewed as being bandwidth inefficient, since we need 16 waveforms to transmit a 4-bit data.

On the other hand, in the proposed method, as shown in Fig. 5.4b, when the same 4-bit data set is partitioned into two subsets, the number of waveforms falls to 8. Similarly, in the conventional method, an 8-bit data set would require $2^8=256$ waveforms, while the proposed method requires only $2\times8=16$ waveforms. This is a substantial reduction of bandwidth indeed [5].

Table 5.1 below shows a comparison between the conventional M-ary signaling and the proposed M-ary signaling for several data lengths. In the conventional method (Col-2, Table 5.1 a k-bit data set requires 2^k waveforms where $k = 1, 2,$ ... Thus the number of waveforms increases rapidly as the length of the data set increases. For these reasons, the conventional method of waveform coding is bandwidth inefficient.

In the proposed method, a k-bit data set requires only $2k$ waveforms where $k = 1, 2, \ldots$(Col-3, Table 5.1). Clearly, the proposed method of waveform coding is bandwidth efficient. Now, our objective is to show that the proposed method of waveform coding applies to bi-orthogonal signaling. We also intend to show that there is a built-in error control mechanism in this scheme. Forward error control coding (FECC) schemes normally used in digital communication systems are not needed in the proposed method. Therefore the proposed method is also cost-effective.

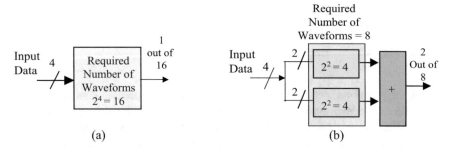

Fig. 5.4 Illustration of waveform coding. (**a**) Conventional method: A 4-bit data set requires 16 waveforms. (**b**) Proposed method: A 4-bit data set partitioned into two data blocks requires only 8 waveforms

Table 5.1 Comparison of bandwidth requirements

	Conventional method	Proposed method	Bandwidth reduction
# Bits (x)	# Waveforms (2^x)	# Waveforms $(2x)$	Factor $(2^x/2x)$
1	2	2	1
2	4	4	1
3	8	6	1.333333333
4	16	8	2
5	32	10	3.2
6	64	12	5.333333333
7	128	14	9.142857143
8	256	16	16
9	512	18	28.44444444
10	1024	20	51.2
11	2048	22	93.09090909
12	4096	24	170.6666667
13	8192	26	315.0769231
14	16384	28	585.1428571
15	32768	30	1092.266667
16	65536	32	2048

5.3   Orthogonal Codes and Antipodal Codes

Orthogonal codes, also known as Walsh codes, were originally developed by J. L. Walsh in 1923 [5]. Walsh codes are well-known for their orthogonal properties. They have been successfully implemented in CDMA for spreading and user ID [6–9]. The use of orthogonal codes for forward error control coding (FECC) has also been investigated by a limited number of authors and has concluded that orthogonal codes do not offer bandwidth efficiency [1]. In this chapter, our goal is to show that orthogonal codes offer error control coding with bandwidth efficiency. We accomplish this by noting that orthogonal codes are binary values and that they have equal number of 1's and zero's. The distance between each orthogonal code is $n/2$, where n is the code length. Since the distance properties are fundamental in error control coding, we show that an n-bit orthogonal code can correct multiple errors with bandwidth efficiency.

5.3.1   *Construction of Orthogonal and Antipodal Codes*

Orthogonal codes are binary valued and can be generated by means of an $N \times N$ Hadamard matrix as follows [10]:

- Construct an $N \times N$ matrix as 4-Quadrants:

N x N Matrix

1st Quadrant	2nd Quadrant
3 rd Quadrant	4th Quadrant

- Keep the 1st, the 2nd, and the 3rd quadrants identical and invert the 4th as follows:

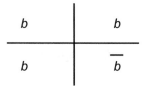

where b is a binary bit which can be either 0 or 1. This process governs the generation of an $N \times N$ Hadamard matrix for N-orthogonal codes with $b=0$ and an $N \times N$ Hadamard matrix for an N-bit antipodal code with $b=1$.

For example, a 2×2 Hadamard matrix generates 2 orthogonal codes and 2 antipodal codes, for a total of 4 bi-orthogonal codes as follows:

Hadamard Matrix	Orthogonal Code block	Antipodal Code block
b b b $\underline{b}$	0 0 0 1	1 1 1 0

where a 2×2 Hadamard matrix generate two orthogonal codes, having two bits each as shown below:

2-bit orthogonal code block	2-bit antipodal code block
0 0	1 1
0 1	1 0

Similarly, a 4 × 4 Hadamard matrix generate 4 orthogonal codes and 4 antipodal codes, for a total of 8 bi-orthogonal codes as follows:

Hadamard Matrix   Orthogonal Code block   Antipodal Code block

Here, we see that a 4 × 4 Hadamard matrix generates 4 orthogonal and 4 antipodal codes, for a total of 8 bi-orthogonal codes as tabulated below:

4-bit orthogonal code block	4-bit antipodal code block
0 0 0 0	1 1 1 1
0 1 0 1	1 0 1 0
0 0 1 1	1 1 0 0
0 1 1 0	1 0 0 1

This principle can be extended to generate n orthogonal codes and n antipodal codes, for a total of $2n$ bi-orthogonal codes. Table below provides a few orthogonal and antipodal code sts for $n=2, 4, 8, 16, 32$, and 64.

Code length (n)	Orthogonal codes (n)	Antipodal codes (n)	Bi-orthogonal codes ($2n$)
2	2	2	4
4	4	4	8
8	8	8	16
16	16	16	32
32	32	32	64
64	64	64	128
:	:	:	:

Fig. 5.5 Bi-orthogonal
code set for $n=8$. An 8-bit
orthogonal code has
8 orthogonal codes and
8-antipodal code for a total
of 16 bi-orthogonal codes

Orthogonal Code	Antipodal Code
0 0 0 0 0 0 0 0	1 1 1 1 1 1 1 1
0 1 0 1 0 1 0 1	1 0 1 0 1 0 1 0
0 0 1 1 0 0 1 1	1 1 0 0 1 1 0 0
0 1 1 0 0 1 1 0	1 0 0 1 1 0 0 1
0 0 0 0 1 1 1 1	1 1 1 1 0 0 0 0
0 1 0 1 1 0 1 0	1 0 1 0 0 1 0 1
0 0 1 1 1 1 0 0	1 1 0 0 0 0 1 1
0 1 1 0 1 0 0 1	1 0 0 1 0 1 1 0

5.3.2   Bi-Orthogonal Codes

In the above, we have established that orthogonal codes are binary valued and have
equal numbers of 1s and 0s. Antipodal codes, on the other hand, are just the inverse
of orthogonal codes. Antipodal codes are also orthogonal among them. Therefore, an
n-bit orthogonal code has n-orthogonal codes and n-antipodal codes, for a total of $2n$
bi-orthogonal codes. For example, an 8-bit orthogonal code has 8 orthogonal codes
and 8 antipodal codes, for a total of 16 bi-orthogonal codes as shown in Fig. 5.5 [11].

Similarly, a 16-bit orthogonal code has 16 orthogonal code and 16-antipodal code
for a total of 32 bi-orthogonal codes, as shown in Fig. 5.6. We will take this
bi-orthogonal code block as an example and examine the error control properties.

5.3.3   Distance Properties of Orthogonal Codes

An n-bit orthogonal code has $n/2$ 1s and $n/2$ 0s, i.e., there are $n/2$ positions where 1's
and 0's differ. Similarly, an n-bit antipodal code has $n/2$ 1s and $n/2$ 0s, i.e., there are
$n/2$ positions where 1's and 0's differ. On the other hand, the distance between an
orthogonal code and an antipodal code is n, where n is the code length. For $n =$
8, these properties can be directly verified from Fig. 5.7, and in Fig. 5.8 where the
distance between any orthogonal code is $8/2 = 4$, while the distance between an
orthogonal code and an antipodal code is 8. This distance property can be used as a
method of error control, as presented in the following section [11].

5.4   Error Control Coding Based on Orthogonal Codes

5.4.1   Error Control Capabilities of Orthogonal Codes

Orthogonal codes are used in CDMA for spreading and user ID. The use of
orthogonal codes for forward error control coding (FECC) has also been investigated

16 Bit Orthogonal Code   16 Bit Antipodal Code

```
0 0 0 0  0 0 0 0  0 0 0 0  0 0 0 0
0 1 0 1  0 1 0 1  0 1 0 1  0 1 0 1
0 0 1 1  0 0 1 1  0 0 1 1  0 0 1 1
0 1 1 0  0 1 1 0  0 1 1 0  0 1 1 0
0 0 0 0  1 1 1 1  0 0 0 0  1 1 1 1
0 1 0 1  1 0 1 0  0 1 0 1  1 0 1 0
0 0 1 1  1 1 0 0  0 0 1 1  1 1 0 0
0 1 1 0  1 0 0 1  0 1 1 0  1 0 0 1

0 0 0 0  0 0 0 0  1 1 1 1  1 1 1 1
0 1 0 1  0 1 0 1  1 0 1 0  1 0 1 0
0 0 1 1  0 0 1 1  1 1 0 0  1 1 0 0
0 1 1 0  0 1 1 0  1 0 0 1  1 0 0 1
0 0 0 0  1 1 1 1  1 1 1 1  0 0 0 0
0 1 0 1  1 0 1 0  1 0 1 0  0 1 0 1
0 0 1 1  1 1 0 0  1 1 0 0  0 0 1 1
0 1 1 0  1 0 0 1  1 0 0 1  0 1 1 0
```

```
1 1 1 1 1 1 1 1 1 1 1 1 1 1 1 1
1 0 1 0 1 0 1 0 1 0 1 0 1 0 1 0
1 1 0 0 1 1 0 0 1 1 0 0 1 1 0 0
1 0 0 1 1 0 0 1 1 0 0 1 1 0 0 1
1 1 1 1 0 0 0 0 1 1 1 1 0 0 0 0
1 0 1 0 0 1 0 1 1 0 1 0 0 1 0 1
1 1 0 0 0 0 1 1 1 1 0 0 0 0 1 1
1 0 0 1 0 1 1 0 1 0 0 1 0 1 1 0
1 1 1 1 1 1 1 1 0 0 0 0 0 0 0 0
1 0 1 0 1 0 1 0 0 1 0 1 0 1 0 1
1 1 0 0 1 1 0 0 0 0 1 1 0 0 1 1
1 0 0 1 1 0 0 1 0 1 1 0 0 1 1 0
1 1 1 1 0 0 0 0 0 0 0 0 1 1 1 1
1 0 1 0 0 1 0 1 0 1 0 1 0 1 1 0 1 0
1 1 0 0 0 0 1 1 0 0 1 1 1 1 0 0
1 0 0 1 0 1 1 0 0 1 1 0 1 1 0 1 0 0 1
```

Fig. 5.6 Bi-orthogonal code set for $n=16$. A 16-bit orthogonal code has 16 orthogonal codes and 16-antipodal code for a total of 32 bi-orthogonal codes

Orthogonal   Distance Properties
Code

C_o	0 0 0 0 0 0 0 0
C_1	0 1 0 1 0 1 0 1
C_2	0 0 1 1 0 0 1 1
C_3	0 1 1 0 0 1 1 0
C_4	0 0 0 0 1 1 1 1
C_5	0 1 0 1 1 0 1 0
C_6	0 0 1 1 1 1 0 0
C_7	0 1 1 0 1 0 0 1

C_o: 0 0 0 0 0 0 0 0
C_1: 0 1 0 1 0 1 0 1
MOD2 ADD: 0 1 0 1 0 1 0 1
Distance Between C_o and C_1 is 4

C_1: 0 1 0 1 0 1 0 1
C_2: 0 0 1 1 0 0 1 1
MOD2 ADD: 0 1 1 0 0 1 1 0
Distance Between C_1 and C_2 is 4

Fig. 5.7 Distance properties of orthogonal codes. For $n=8$, the distance between any orthogonal coded is $8/2=4$. There are $n/2=8/2=4$ positions where 1's and 0's differ

by a limited number of authors. They have concluded that orthogonal codes do not utilize bandwidth efficiently [1]. Our objective in this paper is to investigate the distance properties of orthogonal codes and develop a bandwidth-efficient coded modulation scheme. In the proposed method, orthogonal codes are treated as (n, k) block codes where a k-bit information is represented by a unique n-bit orthogonal

Antipodal
Code

Distance Properties

Co: 1 1 1 1 1 1 1 1

Co	1 1 1 1 1 1 1 1
C1	1 0 1 0 1 0 1 0
C2	1 1 0 0 1 1 0 0
C3	1 0 0 1 1 0 0 1
C4	1 1 1 1 0 0 0 0
C5	1 0 1 0 0 1 0 1
C6	1 1 0 0 0 0 1 1
C7	1 0 0 1 0 1 1 0

C1:1 0 1 0 1 0 1 0

MOD2 ADD:1 0 1 0 1 0 1 0
Distance Between Co and C1 is 4

C1: 1 0 1 0 1 0 1 0
C2: 1 1 0 0 1 1 0 0

MOD2 ADD: 0 1 1 0 0 1 1 0
Distance Between Co and C1 is 4

Fig. 5.8 Distance properties of antipodal codes. For $n = 8$, the distance between any antipodal coded is $8/2 = 4$. There are $n/2 = 8/2 = 4$ positions where 1's and 0's differ

code ($k < n$). We examine this by noting that an n-bit orthogonal code has $n/2$ 1s and $n/2$ 0s, i.e., there are $n/2$ positions where 1s and 0s differ. Therefore, the distance between two orthogonal codes is also $n/2$, where each orthogonal code generates a zero parity bit. These properties are exploited to detect and correct errors with bandwidth efficiency. We show that an n-bit orthogonal code can correct t errors where $t = n/4 - 1$, n being the code length. A measure of coding gain is then obtained by comparing the word error with coding to the word error without coding. Construction of rate 1 orthogonal coded modulation schemes, using an 8-bit orthogonal code, is presented to illustrate the concept [2–4, 9, 11].

In order to examine the error control properties of orthogonal codes, we note that an n-bit orthogonal code has $n/2$ 1s and $n/2$ 0s, i.e., there are $n/2$ positions where 1s and 0s differ. Therefore, the distance between two orthogonal codes is $d = n/2$. This distance property can be used to detect an impaired received code by setting a threshold midway between two orthogonal codes as shown in Fig. 5.9, where the received coded is shown as a dotted line. This is given by the following equation:

$$d_{th} = \frac{n}{4} \tag{5.1}$$

where n is the code length, and d_{th} is the threshold, which is midway between two orthogonal codes. Therefore, for the 8-bit orthogonal code (Fig. 5.3), we have $d_{th} = 8/4 = 2$. This mechanism offers a decision process, where the incoming impaired orthogonal code is examined for correlation with the neighboring codes for a possible match.

The acceptance criterion for a valid code is that an n-bit comparison must yield a good autocorrelation value; otherwise, a false detection will occur. The following

Fig. 5.9 Decoding principle. The received code is compared to a lookup table for a possible match

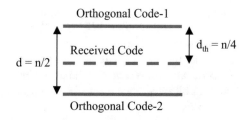

Table 5.2 Orthogonal codes and the corresponding error correction capabilities

Code length n	Number of errors corrected t
8	1
16	3
32	7
64	15

correlation process governs this where an impaired orthogonal code is compared with a pair of n-bit orthogonal codes to yield:

$$R(x, y) = \sum_{i=1}^{n} x_i y_i \geq (n - d_{th}) + 1 \qquad (5.2)$$

where $R(x, y)$ is the autocorrelation function, n is the code length, d_{th} is the threshold, as defined earlier. Since the threshold (d_{th}) is in the midway between two valid codes, an additional 1-bit offset is added to Eq. (5.4) for reliable detection. The number of errors that can be corrected by means of this process can be estimated by combining Eqs. (5.3) and (5.4), yielding:

$$t = n - R(x, y) = \frac{n}{4} - 1 \qquad (5.3)$$

In the above equation, t is the number of errors that can be corrected by means of an n-bit orthogonal code. For example, a single error-correcting orthogonal code can be constructed by means of an 8-bit orthogonal code ($n = 8$). Similarly, a three-error-correcting, orthogonal code can be constructed by means of a 16-bit orthogonal code ($n = 16$), and so on. Table 5.2 shows a few orthogonal codes and the corresponding error-correcting capabilities:

5.4.2   Error Performance and Coding Gain

In the previous section, we have established that an n-bit orthogonal code can correct t errors where $t = (n/4) - 1$, n being the code length. A measure of coding gain is then obtained by comparing the word error without coding WER(U) to the word error with coding WER(C). We examine this by means of the following analytical means [12]:

Let a k-bit data set be represented by an n-bit orthogonal code, where $n > k$. Then the code rate will be k/n, and the coded bit rate will be $R_c = (n/k)R_b$, where R_b is the uncoded bit rate. Since $n > k$, the coded bit rate R_c will be greater than the uncoded bit rate R_b ($R_c > R_b$). Consequently, the coded bit energy E_c will be less than the uncoded bit energy E_b ($E_c < E_b$). If S is the transmit carrier power, then the uncoded bit energy (E_b) and the coded bit energy (E_c) will be:

$$E_b = \frac{S}{R_b} \tag{5.4}$$

$$E_c = E_b \left(\frac{k}{n}\right) = \left(\frac{S}{R_b}\right)\left(\frac{k}{n}\right) \tag{5.5}$$

With orthogonal on-off keying modulation and noncoherent detection, the uncoded bit error probability P_{eu} and the coded bit error probability P_{ec} over AWGN (additive White Gaussian noise) channel without fading are given by:

$$P_{eu} \approx \frac{1}{2} Exp\left(\frac{-E_b}{2N_0}\right) = \frac{1}{2} Exp\left(\frac{-S}{2R_bN_0}\right) \tag{5.6}$$

$$P_{ec} \approx \frac{1}{2} Exp\left(\frac{-E_c}{2N_0}\right) = \frac{1}{2} Exp\left[\left(\frac{-S}{2R_bN_0}\right)\left(\frac{k}{n}\right)\right] \tag{5.7}$$

where E_b/N_o is the energy per bit to noise spectral density. E_b/N_o is related to S/N (signal to noise) ratio, also known as "SNR" as follows:

$$E_b/N_0 = \left(\frac{S}{N_0R_b}\right) = \left(\frac{S}{N}\right)\left(\frac{W}{R_b}\right) \tag{5.8}$$

S/N is the ratio of average signal power to average noise power, where $N = N_oW$, $W =$ signal bandwidth. E_b/N_0 is the normalized measure of the energy per symbol to noise power spectral density. The parameter E_b/N_0 is generally used to estimate the bit error rate (BER) performance of different digital modulation schemes.

Since $n > k$, the coded bit error will be more than the uncoded bit error. However, it still remains to be seen whether there is a net gain in word error rate due to coding. This can be achieved by comparing the uncoded word error rate WER(U) with the coded word error rate WER(C). These word error rates over AWGN channel without fading are given by:

$$\text{WER}\,(U) = 1 - (1 - P_{eu})^k \tag{5.9}$$

$$\text{WER}\,(C) = \sum_{i=t+1}^{n} \binom{n}{i} P_{ec}(1 - P_{ec})^{n-i} \tag{5.10}$$

where P_{eu} is the uncoded bit error rate, Pec is the coded bit error rate, and t is the maximum errors corrected by the code. For rate ½ orthogonal codes, the word error rate (WER) for orthogonal on-off-keying (O3K) modulation were calculated for various code lengths and plotted in the graph as shown in Fig. 5.10 [13]. The uncoded BER is also plotted for comparison.

Coding gain is the difference in E_b/N_0 between the uncoded and the coded word errors. Notice that at least 3–7 dB coding gains are available in this example. We also note that coding gain increases for longer codes. From these results, we conclude that orthogonal codes offer coding gain.

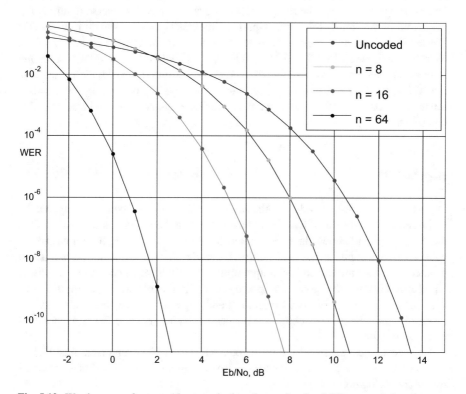

Fig. 5.10 Word error performance due to a single orthogonal code of different code lengths

5.5   Wave Coding Based on Orthogonal Codes

Orthogonal codes are essentially (n, k) block codes, where a k-bit data set is represented by a unique n-bit orthogonal code $(k < n)$. We illustrate this by means of an 8-bit orthogonal code, having 8-orthogonal and 8-antipodal codes for a total of 16 bi-orthogonal codes. We assume that an n-bit orthogonal code can be treated as an (n, k) block code. We now show that code rates such as rate ½, rate ¾, and rate 1 are indeed available out of orthogonal codes. The principle is presented below.

5.5.1   Construction of Rate 1/2 Waveform Coding

Encoder

A rate 1/2 orthogonal coded modulation with an 8-bit orthogonal code, having 16 bi-orthogonal codes, $(m=16, n = 8)$, can be constructed by inverse multiplexing the incoming traffic, R_b (b/s), into 4-parallel streams $(k = 4)$ as shown in Fig. 5.11. These bit streams, now reduced in speed to $R_b/4$ (b/s), are used to address 16 8-bit bi-orthogonal codes, stored in a 16×8 ROM. The output of the ROM is a unique 8-bit orthogonal code, which is modulated and transmitted through a channel. The modulated waveform is in orthogonal space. The code rate is given by $r=4/8=1/2$. Since there is only one 8-bit orthogonal waveform, only one error can be corrected in this scheme.

Decoder

Decoding is a process of code correlation. In this process, the receiver compares the incoming impaired data with the actual data stored in the lookup table for a possible match. Figure 5.12 shows the lookup table at the receiver. Notice that for each 4-bit data, there is unique orthogonal (antipodal) code. According to the transmission protocol, the transmitter sends a unique orthogonal (antipodal) code to the receiver. Upon receiving an orthogonal (antipodal) code, the receiver validates the received data pattern by means of code correlation and appends a correlation value for the received data. The process continues for each entry to generate the corresponding correlation value.

The acceptance criterion for a valid code is that an n-bit comparison must yield a good autocorrelation value; otherwise, a false detection will occur. . This is governed by the following correlation value:

$$t \le (n/4) - 1$$

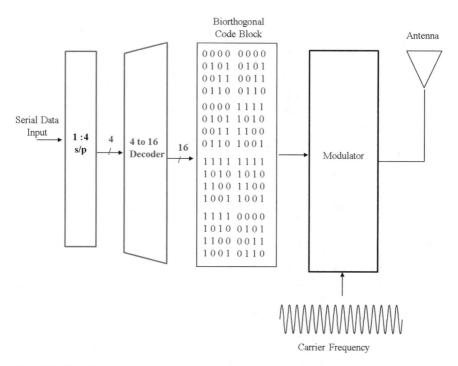

Fig. 5.11 Rate 1/2 orthogonal coded modulation with $n = 8$

where n is the code length, and t is the number of errors that can be corrected by an n-bit orthogonal code.

Let's examine the correlation process using the following example:

The input bit pattern: $k = 0\ 0\ 1\ 1$
Encoded Transmit code: $n = 0\ 1\ 1\ 0\ 0\ 1\ 1\ 0$
Received impaired code: $n^* = 0\ 1\ 1\ 0\ 0\ 1\ 1\ \mathbf{1}$ (the last bit is in error)

Notice that the last bit is in error, identified in bold. Now, let's determine how the receiver recovers the correct data by means of code correlation. The correlation process is described below:

Test 1

This test compares the received data with the 1st row of data stored in the lookup table and counts the number of positions it does not match. This is accomplished by MOD2 operation (EXOR operation). The result is presented below:

Received impaired code: 0 1 1 0 0 1 1 1
1st row of code in the lookup table: 0 0 0 0 0 0 0 0
Mod-2 Add: 0 1 1 0 0 1 1 1
Correlation Value = 5 (count the number of 1's in MOD2 Add)
Verdict: No match, Continue search

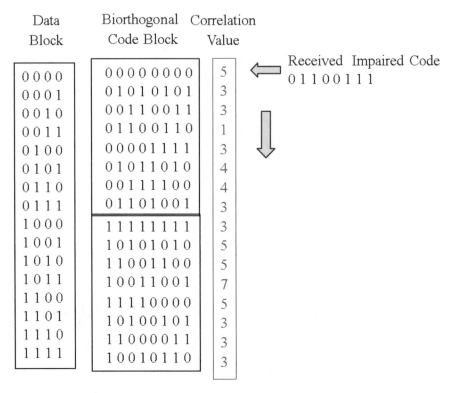

Fig. 5.12 Correlation receiver for rate ½ bi-orthogonal coding. Upon receiving an impaired code, the receiver compares it with each entry in the code block and appends a correlation value for each comparison. A valid code is declared when the closest approximation is achieved. For rate ½, $n=8$, this value is 1, and the corresponding data is 0011

Test 2

This test compares the received data with the 2nd row of data stored in the lookup table and counts the number of positions it does not match. This is accomplished by MOD2 operation (EXOR operation). The result is presented below:

Received impaired code: 0 1 1 0 0 1 1 1
2nd row of code in the lookup table: 0 1 0 1 0 1 0 1
Mod-2 Add: 0 01 1 0 01 0
Correlation Value = 3
Verdict: No match, Continue search

In a similar manner, we can determine the remaining correlation values as depicted in Fig. 5.12. Notice that the lowest correlation value is 1, which is the valid code. Therefore, the corresponding 4 bit data is 0 0 1 1, which has been transmitted.

5.5.2   *Construction of Rate 3/4 Waveform Coding*

Encoder

A rate ¾ orthogonal coded modulation with an 8-bit orthogonal code, having 16 bi-orthogonal codes, ($m=16$, $n = 8$), can be constructed by inverse multiplexing the incoming traffic, R_b (b/s), into 6-parallel streams ($k = 6$) as shown in Fig. 5.13. These bit streams, now reduced in speed to $R_b/6$ (b/s), are partitioned into two 8×3 data blocks. The first 8×3 data block maps the 8×8 orthogonal code block, and the next 8×3 data block maps the 8×8 antipodal code block. These code blocks are stored in two 8 × 8 ROMs. The output of each ROM is a unique 8-bit orthogonal/ antipodal code, which are modulated by means of the respective modulator using the same carrier frequency.

The code rate is given by $r=6/8=3/4$. Since there are two orthogonal waveforms (one orthogonal and one antipodal), the number of errors that can be corrected is given by 2. Moreover, the bandwidth is also reduced.

Decoder

Decoding is a correlation process similar to the one presented earlier. Let's examine the correlation process using the following example:

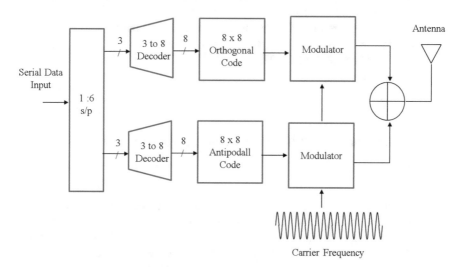

Fig. 5.13 Rate 3/4 orthogonal coded modulation with $n = 8$

For data block 1:

Input bit pattern: $k1 = 0\ 1\ 1$
Encoded transmit code: $n\ 1 = 0\ 1\ 1\ 0\ 0\ 1\ 1\ 0$
Received impaired code: $n1^* = 0\ 1\ 1\ 0\ 0\ 1\ 1\ \mathbf{1}$ (the last bit is in error)

   Notice that the last bit is in error, identified in bold.

For data block 2:

Input bit pattern: $k2 = 1\ 0\ 0$
Encoded transmit code: $n\ 2 = 1\ 1\ 1\ 1\ 0\ 0\ 0\ 0$
Received impaired code: $n1^* = \mathbf{0}\ 1\ 1\ 1\ 0\ 0\ 0\ \mathbf{1}$ (the first bit is in error)

   Notice that the first bit is in error, identified in bold.
   Figure 5.14 shows the correlation receiver for rate 3/4 bi-orthogonal coding. Upon receiving an impaired code, the receiver compares it with each entry in the code block and appends a correlation value for each comparison. A valid code is declared when the closest approximation is achieved. For rate 3/4 coding with $n=8$, there are two code blocks, the minimum correlation value is 1 for an orthogonal code and 1 for an antipodal code. The corresponding data is 011 and 100, respectively.
   The correlation process for rate $r = \frac{3}{4}$ with code length $n = 8$ is described below:

- Upon receiving an impaired code, the receiver compares it with each entry in the code block and appends a correlation value for each comparison.
- A valid code is declared when the closest approximation is achieved.
- For rate 3/4 coding with $n=8$, there are two code blocks, the minimum correlation value is 1 for an orthogonal code and 1 for an antipodal code.
- The corresponding data is 011 and 100, respectively.
- Number of errors corrected is 2.
- Code rate is given by $r == \frac{3}{4}$.

5.5.3 Construction of Rate 1 Waveform Coding

Encoder

A rate 1orthogonal coded modulation with an 8-bit orthogonal code, having 16 bi-orthogonal codes, ($m=16$, $n = 8$), can be constructed by inverse multiplexing the incoming traffic, R_b (b/s), into 8-parallel streams ($k = 8$) as shown in Fig. 5.15.
   The bit streams, now reduced in speed to $R_b/8$ (b/s), are partitioned into four data blocks. Each data block is mapped into a 4×8 code block. These code blocks are stored in four 4×8 ROMs as depicted in the figure. The output of each ROM is a unique 8-bit orthogonal/antipodal code, which is modulated by the respective

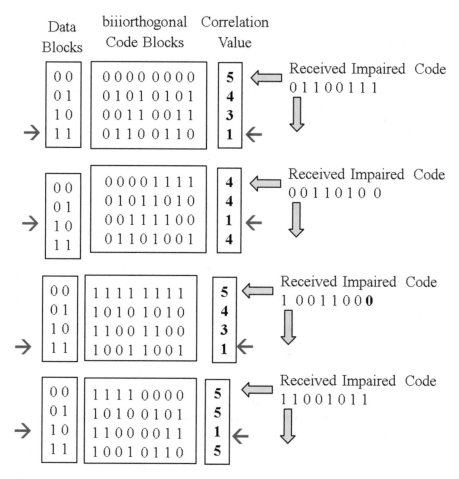

Fig. 5.14 Correlation receiver for rate 3/4 bi-orthogonal coding

modulator using the same carrier frequency. The modulated waveforms are in orthogonal space and have fewer errors. The code rate is given by $r=8/8=1$. This is achieved without bandwidth expansion. Since there are four orthogonal waveforms, the number of errors that can be corrected is 4.

Decoder

Figure 5.16 displays the correlation receiver for rate 1 bi-orthogonal coding with $n=8$ orthogonal codes. Once again, the decoding process is similar to the one presented earlier. Notice that the entire bi-orthogonal code block is partitioned into four code blocks. Each code block represents a data block as shown in the figure.

16 Bit Orthogonal Code	16 Bit Antipodal Code
0000 0000 0000 0000	1111 1111 1111 1111
0101 0101 0101 0101	1010 1010 1010 1010
0011 0011 0011 0011	1100 1100 1100 1100
0110 0110 0110 0110	1001 1001 1001 1001
0000 1111 0000 1111	1111 0000 1111 0000
0101 1010 0101 1010	1010 0101 1010 0101
0011 1100 0011 1100	1100 0011 1100 0011
0110 1001 0110 1001	1001 0110 1001 0110
0000 0000 1111 1111	1111 1111 0000 0000
0101 0101 1010 1010	1010 1010 0101 0101
0011 0011 1100 1100	1100 1100 0011 0011
0110 0110 1001 1001	1001 1001 0110 0110
0000 1111 1111 0000	1111 0000 0000 1111
0101 1010 1010 0101	1010 0101 0101 1010
0011 1100 1100 0011	1100 0011 0011 1100
0110 1001 1001 0110	1001 0110 0110 1001

Fig. 5.15 Rate 1 orthogonal coded modulation with $n = 8$

- Upon receiving an impaired code, the receiver compares it with each entry in the code block and appends a correlation value for each comparison.
- A valid code is declared when the closest approximation is achieved.
- As can be seen, the minimum correlation value in each block is 1 as depicted in the figure.
- The corresponding data is 1 1, 1 0, 1 1, and 1 0, respectively.

Since there are four orthogonal waveforms (two orthogonal and two antipodal), the number of errors that can be corrected is given by $4 \times 1 = 4$. Moreover, the bandwidth is further reduced. The result is summarized below:

- Code rate: $r = 8/8 = 1$
- Number of errors corrected: $4t = 4[(n/4) - 1] = 4(8/4) - 1] = 4$

5.6   Higher-Order Orthogonal Waveform Coding Using 16-Bit Orthogonal Code

A 16-bit orthogonal code has 16 orthogonal codes and 16 antipodal codes, for a total of 32 bi-orthogonal codes as shown in Fig. 5.17. Notice that the distance between any orthogonal codes is $n/2 = 16/2 = 8$. Since the distance properties are fundamental

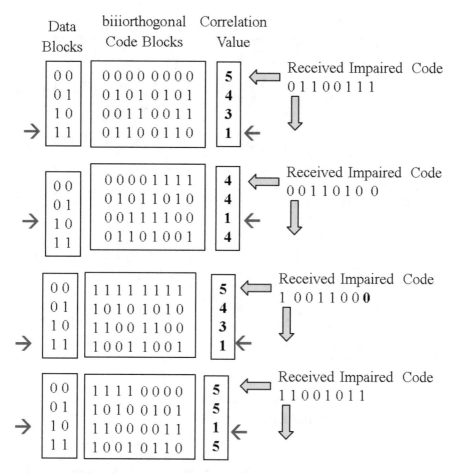

Fig. 5.16 Correlation receiver for rate 1 bi-orthogonal coding with $n=8$. Upon receiving an impaired code, the receiver compares it with each entry in the code block and appends a correlation value for each comparison. A valid code is declared when the closest approximation is achieved. For rate 1 coding with $n=8$, there are four code blocks. The minimum correlation value in each block is 1.The corresponding data is 11, 10, 1 1, and 1 0, respectively

in error control coding, the $n=16$ bit code can correct more errors. The number of errors that can be corrected by each of these codes, is given by:

$$t = (n/4) - 1 = (16/4) - 1 = 3 \qquad (5.11)$$

where $n = 16$ is the code length. These bi-orthogonal codes can be used to realize a variety of waveform coding with bandwidth efficiency. The construction of these code blocks is briefly presented in the following sections.

16 Bit Orthogonal Code

16 Bit Antipodal Code

```
0 0 0 0  0 0 0 0  0 0 0 0  0 0 0 0
0 1 0 1  0 1 0 1  0 1 0 1  0 1 0 1
0 0 1 1  0 0 1 1  0 0 1 1  0 0 1 1
0 1 1 0  0 1 1 0  0 1 1 0  0 1 1 0
0 0 0 0  1 1 1 1  0 0 0 0  1 1 1 1
0 1 0 1  1 0 1 0  0 1 0 1  1 0 1 0
0 0 1 1  1 1 0 0  0 0 1 1  1 1 0 0
0 1 1 0  1 0 0 1  0 1 1 0  1 0 0 1
0 0 0 0  0 0 0 0  1 1 1 1  1 1 1 1
0 1 0 1  0 1 0 1  1 0 1 0  1 0 1 0
0 0 1 1  0 0 1 1  1 1 0 0  1 1 0 0
0 1 1 0  0 1 1 0  1 0 0 1  1 0 0 1
0 0 0 0  1 1 1 1  1 1 1 1  0 0 0 0
0 1 0 1  1 0 1 0  1 0 1 0  0 1 0 1
0 0 1 1  1 1 0 0  1 1 0 0  0 0 1 1
0 1 1 0  1 0 0 1  1 0 0 1  0 1 1 0
```

```
1 1 1 1 1 1 1 1 1 1 1 1 1 1 1 1
1 0 1 0 1 0 1 0 1 0 1 0 1 0 1 0
1 1 0 0 1 1 0 0 1 1 0 0 1 1 0 0
1 0 0 1 1 0 0 1 1 0 0 1 1 0 0 1
1 1 1 1 0 0 0 0 1 1 1 1 0 0 0 0
1 0 1 0 0 1 0 1 1 0 1 0 0 1 0 1
1 1 0 0 0 0 1 1 1 1 0 0 0 0 1 1
1 0 0 1 0 1 1 0 1 0 0 1 0 1 1 0
1 1 1 1 1 1 1 1 0 0 0 0 0 0 0 0
1 0 1 0 1 0 1 0 0 1 0 1 0 1 0 1
1 1 0 0 1 1 0 0 0 0 1 1 0 0 1 1
1 0 0 1 1 0 0 1 0 1 1 0 0 1 1 0
1 1 1 1 0 0 0 0 0 0 0 0 1 1 1 1
1 0 1 0 0 1 0 1 0 1 0 1 1 0 1 0
1 1 0 0 0 0 1 1 0 0 1 1 1 1 0 0
1 0 0 1 0 1 1 0 0 1 1 0 1 1 0 1
```

Fig. 5.17 16-bit bi-orthogonal code set having 16 orthogonal codes and 16 antipodal codes for a total of 32 bi-orthogonal codes

5.6.1 Rate 5/16 Orthogonal Waveform Coding Based on n =16 Orthogonal Code

A rate 5/16 orthogonal coded modulation with a 16-bit orthogonal code, having 32 bi-orthogonal codes, ($m = 32$, $n = 16$), can be constructed by inverse multiplexing the incoming traffic, R_b (b/s), into 5-parallel streams ($k = 5$) as shown in Fig. 5.18. These bit streams, now reduced in speed by a factor of 5, are used to address 32, 16-bit bi-orthogonal codes, stored in a 32 × 16 ROM. The output of the ROM is a unique 16-bit orthogonal (antipodal) code, which is modulated and transmitted through a channel. The modulated waveform is in orthogonal space.

Figure 5.18 also illustrates the correlation receiver for rate 5/16 bi-orthogonal coding based on n=16 orthogonal codes. Notice that the entire 32×5 data block is mapped into the entire 32×16 bi-orthogonal code block. Each 16-bit orthogonal/antipodal code represents a 5-bit data pattern as shown in the figure.

The correlation process is as follows:

- Upon receiving an impaired code, the receiver compares it with each entry in the code block and appends a correlation value for each comparison.
- A valid code is declared when the closest approximation is achieved.
- As can be seen, the minimum correlation value in the entire block is 3, which identifies the valid code and the corresponding data as shown in the figure.

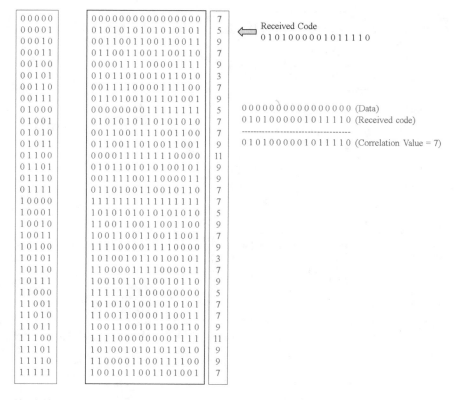

Fig. 5.18 Rate 5/16 bi-orthogonal coding based on $n=16$ orthogonal code

The code rate (r) and the number of errors (t) that can be corrected are as follows:

- Code rate: $r = 5/16$
- Number of errors corrected: $t = (n/4)-1 = (16/4)-1= 3$

Since the code rate is 5/16, this scheme is bandwidth inefficient.

5.6.2 Rate ½ Orthogonal Waveform Coding Based on n =16 Orthogonal Code

A rate ½ orthogonal coded modulation with a 16-bit orthogonal code, having 32 bi-orthogonal codes, ($m=32$, $n = 16$), can be constructed by inverse multiplexing the incoming traffic, R_b (b/s), into 8-parallel streams ($k = 8$), as shown in Fig. 5.19. These bit streams, now reduced in speed by a factor of 8, are partitioned into two data blocks, 4-bits per subset. Each 4-bit subset is used to address eight 16-bit orthogonal codes. These codes are stored in two 16×16 ROMs. The output of each ROM is a

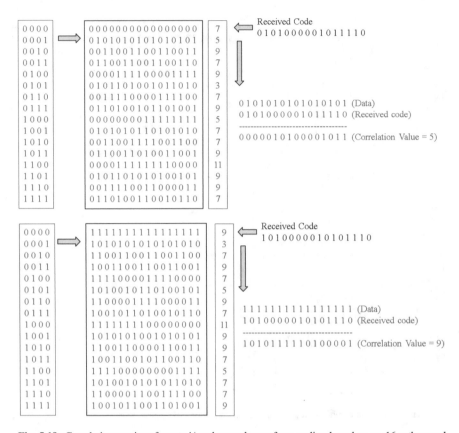

Fig. 5.19 Correlation receiver for rate ½ orthogonal waveform coding based on $n=16$ orthogonal codes

unique 16-bit orthogonal code, which is modulated by the respective modulator and transmitted through a channel. The modulated waveforms are in orthogonal space.

Figure 5.19 also displays the correlation receiver for rate ½ bi-orthogonal coding based on $n=16$ orthogonal codes. In this scheme, both the data and the bi-orthogonal code blocks are partitioned into two blocks as shown in the figure. Each block corrects 3 errors for a total of 6 errors.

Data   Biorthogonal Code Blocks Correlation

The correlation process is as follows:

- Upon receiving an impaired code, the receiver compares it with each entry in the orthogonal code block and appends a correlation value for each comparison.
- The process is similar for the antipodal code block.
- In each block, a valid code is declared when the closest approximation is achieved.
- As can be seen, the minimum correlation value in each block is 3, which identifies the valid code and the corresponding data as shown in the figure.

- The total number of errors that can be corrected by this scheme is given by 6.

Since there are two orthogonal waveforms (one orthogonal and one antipodal), the number of errors that can be corrected is given by $2\times3=6$. Moreover, the bandwidth is also reduced. The result is summarized below:

- Code rate: $r = 8/16 = 1/2$
- Number of errors corrected: $2t=2[(n/4)-1]=2(16/4)-1]=6$

5.6.3 Rate ¾ Orthogonal Waveform Coding Using n=16 Orthogonal Code

A rate ¾ orthogonal coded modulation with a 16-bit orthogonal code, having 32 bi-orthogonal codes, $(m=32, n = 16)$, can be constructed by inverse multiplexing the incoming traffic into 12-parallel streams ($k = 12$) as shown in Fig. 5.20. These bit streams, now reduced in speed by a factor of 12, are partitioned into four data blocks, 3-bits per subset. Each 3-bit subset is used to address eight 16-bit orthogonal codes. These codes are stored in four 8×16 ROMs. The output of each ROM is a unique 16-bit orthogonal (antipodal) code, which is modulated by the respective modulator and transmitted through a channel. The modulated waveforms are in orthogonal space.

Figure 5.20 also displays the correlation receiver for rate 3/4 bi-orthogonal coding based on $n=16$ orthogonal codes. In this scheme, both the data and the bi-orthogonal code blocks are partitioned into four blocks as shown in the figure. Each block corrects 3 errors for a total of 12 errors.

The correlation process is as follows:

- Upon receiving an impaired code, the receiver compares it with each entry in the orthogonal code block and appends a correlation value for each comparison.
- The process is similar for the remaining code blocks.
- In each block, a valid code is declared when the closest approximation is achieved.
- As can be seen, the minimum correlation value in each block is 3, which identifies the valid code and the corresponding data as shown in the figure.

Since there are four orthogonal waveforms (two orthogonal and two antipodal), the number of errors that can be corrected is given by $4\times3=12$. Moreover, the bandwidth is further reduced. The result is summarized below:

- Code rate: $r =12/16=3/4$
- Number of errors corrected: $4t=4[(n/4)-1]=4(16/4)-1]=12$

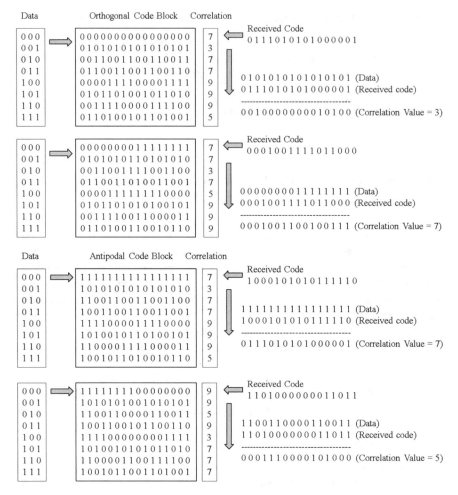

Fig. 5.20 Correlation receiver for rate ¾ orthogonal waveform coding based on $n=16$ orthogonal codes

5.6.4 Rate 1 Orthogonal Waveform Coding Based on n=16 Orthogonal Code

A rate 1 orthogonal coded modulation with a 16-bit orthogonal code, having 32 bi-orthogonal codes, $(m=32, n=16)$, can be constructed by inverse multiplexing the incoming traffic, R_b (b/s), into 16-parallel streams ($k=16$) as shown in Fig. 5.21. These bit streams, now reduced in speed by a factor of 16, are partitioned into 8 data blocks, 2-bits per subset. Each 2-bit subset is used to address four 16-bit orthogonal codes. These codes are stored in eight 4×16 ROMs. The output of each ROM is a unique 16-bit orthogonal (antipodal) code, which is modulated by the respective modulator and transmitted through a channel.

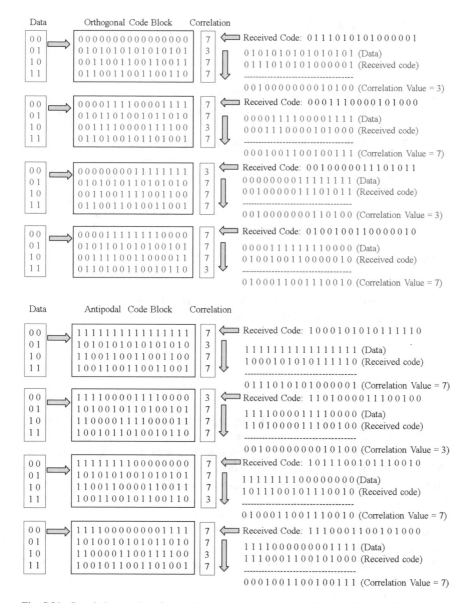

Fig. 5.21 Correlation receiver for rate 1 orthogonal waveform coding based on $n=16$ orthogonal codes

Figure 5.21 also displays the correlation receiver for rate 1 bi-orthogonal coding based on $n=16$ orthogonal codes. In this scheme, both the data and the bi-orthogonal code blocks are partitioned into eight blocks as shown in the figure. Each block corrects 3 errors for a total of 24 errors.

The correlation process is as follows:

- Upon receiving an impaired code, the receiver compares it with each entry in the orthogonal /antipodal code block and appends a correlation value for each comparison.
- The process is carried out for each block for a total of eight blocks.
- In each block, a valid code is declared when the closest approximation is achieved.
- As can be seen, the minimum correlation value in each block is 3, which identifies the valid code and the corresponding data as shown in the figure.
- Since there are eight blocks, the total number of errors that can be corrected by this scheme is given by $3\times8=24$.

Since there are eight orthogonal waveforms (four orthogonal and four antipodal), the number of errors that can be corrected is given by $8\times3=24$. Moreover, the bandwidth is further reduced.

The result is summarized below:

- Code rate: $r=16/16=1$.
- Number of errors corrected: $8t=8[(n/4)-1]=8(16/4)-1]=24$.
- This is achieved without bandwidth expansion.

5.7   Waveform Capacity

Waveform coding is a technique where multiple parallel data streams from a single user are used to modulate the same carrier frequency. In this scheme, the modulated carrier frequencies are in orthogonal space and have a unique noise signature. Therefore, waveform capacity can be defined as the number of modulated waveforms that can be combined without interfering each other. This is conceptually shown in Fig. 5.22, where N waveforms share the same transmission bandwidth W. Each waveform contributes noise before data recovery.

Our objective is to determine the number N. that can be supported in a given bandwidth. This is similar to CDMA capacity [6] as presented below.

From the circuit theory, we know that the power delivered into a load is the rate of change of energy, which is given by:

$$P = \frac{dE}{dt} \tag{5.12}$$

where

- $P =$ power
- $E =$ energy

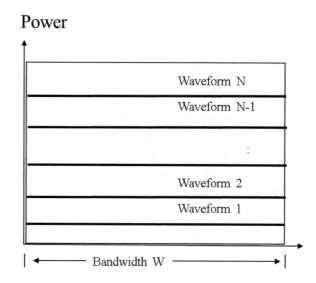

Fig. 5.22 Illustration of N waveforms sharing the same transmission bandwidth W. Each waveform represents a noise source before data recovery

On the other hand, in digital communication, we define power and energy as follows:

$$C = \frac{E_b}{T} = E_b R_b \qquad (5.13)$$

where

- C = carrier power
- E_B = energy per bit
- T- = bit duration
- R_b = bit rate (b/s)

Now, let's define:

- W = bandwidth
- I = total interference (noise) due to multiple users

Then the noise density N_o can be written as:

$$N_0 = \frac{I}{W}$$
or   (5.14)
$$I = N_0 \times W$$

From the above equations, we can express the carrier to interference ratio as follows:

$$\frac{C}{I} = \frac{R_b \times E_b}{N_o \times W} \tag{5.15}$$

where

- $E_B =$ energy per bit
- $R_b =$ bit rate
- $N_o=$ noise density (also called "noise spectral density")
- $W =$ transmission bandwidth

In a correlation receiver, the interference is due to all users except the one which is being recovered by means of code correlation as depicted in Fig. 5.23. Therefore, we can write:

$$\begin{aligned} I &= C(N-1) \\ \frac{C}{I} &= \frac{1}{N-1} \end{aligned} \tag{5.16}$$

From Eqs. (5.15) and (5.16), we get:

$$\frac{1}{N-1} = \frac{R_b \times E_b}{N_o \times W} \tag{5.17}$$

Solving for the capacity N, we obtain:

$$N = 1 + \frac{W/R_b}{E_b/N_o} \tag{5.18}$$

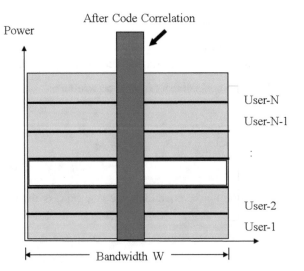

Fig. 5.23 Correlation receiver outcome. One out of N signals is correlated, and its signal strength is the highest

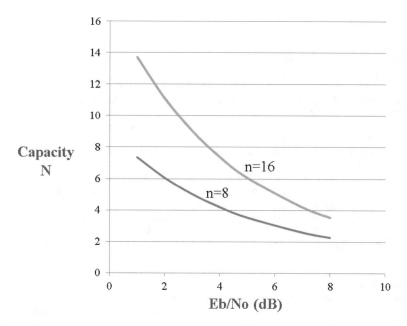

Fig. 5.24 Waveform capacity as a function of E_B/N_o in dB

In the above equation, N is the waveform capacity. This is the number of waveforms that can be combined for a given code length, E_B/No is the energy per bit to noise ratio, and W/R_b is the data spreading factor. For $n=8$, this value is 8, and for $n=16$, this value is 16. Figure 5.24 shows the waveform capacity as a function E_B/No.

5.8   Conclusions

- This chapter presents a method of waveform coding, based on orthogonal codes.
- In the proposed method, the high-speed data stream is inverse multiplexed into several parallel streams.
- These parallel streams, now reduced in speed, are partitioned into several blocks and mapped into blocks of bi-orthogonal codes.
- A bank of identical modulators are used to modulate the coded bit streams using the same carrier frequency.
- Construction of rate ½, rate ¾, and rate 1 waveform coding schemes are presented to illustrate the concept.
- It is also shown that there is a built-in error control mechanism in this scheme.
- The proposed method is bandwidth efficient.

References

1. B. Sklar, *Digital Communications Fundamentals and Applications* (Prentice Hall, Upper Saddle River, 1988)
2. S. Faruque, *Battlefield Wideband Transceivers Based on Combined N-ary Orthogonal Signaling and M-ary PSK Modulation*, SPIE Proceedings, Vol. 3709 "Digitization of the Battle Space 1V", pp. 123–128, 1999
3. S. Faruque, et al., *Broadband Wireless Access Based on Code Division Parallel Access*, US Patent No. 6208615, March 27, 2001
4. S. Faruque, et al., *Bi-Orthogonal Code Division Multiple Access System*, US Patent No. 6198719, March 6, 2001
5. J.L. Walsh, A closed set of normal orthogonal functions. Am. J. Math. **45**(1), 5–24 (1923)
6. IS-95 *Mobile Station – Base Station Compatibility Standard for dual Mode Wide band Spread Spectrum Cellular Systems*, TR 55, PN-3115, March 15, 1993
7. S.C. Yang, *CDMA RF System Engineering* (Artech House Inc, Boston, 1998)
8. V.K. Garg, *IS-95 CDMA and cdma 2000* (Prentice Hall, 1999)
9. S. Faruque, *Code Division Multiple Access Cable Modem*, US Patent No. 6647059, November, 2003
10. S. Faruque, *Cellular Mobile Systems Engineering* (Artec House Inc., Norwood, 1996) ISBN: 0-89006-518-7
11. S. Faruque, *Broadband Communications Based On Code Division Parallel Access (CDPA)*, International Engineering Consortium (IEC), Annual Review of Communications, Vol. 57, ISBN: 1-931695-28-8, November 2004
12. G. Ungerboeck, Channel coding with multilevel/multiphase signals. IEEE Trans. Inf. Theory **IT28**, 55–67 (1982)
13. S. Faruque, S. Faruque, W. Semke, *Orthogonal On-Off Keying (O3K) for Free-Space Laser Communications With Ambient Light Compensation*, SPIE Optical Engineering, Volume 52, Issue 9, Lasers, Fiber Optics, and Communications, 2013 (Published)

Chapter 6
Introduction to Modulation

Topics

- Background
- Modulation by Analog Signal
- AM and FM Bandwidth at a Glance
- Modulation by Digital Signal
- ASK, FSK and PSK Bandwidth at a Glance

6.1 Background

Modulation is a technique that changes the characteristics of the carrier frequency in accordance to the input signal. Figure 6.1 shows the conceptual block diagram of a modern wireless communication system, where the modulation block is shown in the inset of the dotted block. As shown in the figure, modulation is performed at the transmit side, and demodulation is performed at the receive side. This is the final stage of any radio communication system. The preceding two stages have been discussed elaborately in my previous book in this series [1, 2].

The output signal of the modulator, referred to as the modulated signal, is fed into the antenna for propagation. Antenna is a reciprocal device that transmits and receives the modulated carrier frequency. The size of the antenna depends on the wavelength (λ) of the sinusoidal wave where

$$\lambda = c/f \text{ meter}$$

c = velocity of light = 3×10^8 m/s
f = frequency of the sinusoidal wave, also known as "carrier frequency"

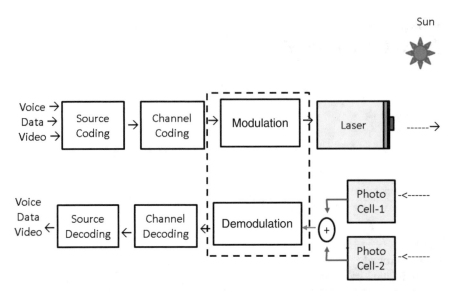

Fig. 6.1 Block diagram of a full-duplex laser communication with ambient light compensation. The modulation block is shown in the inset of the dotted block

In general, a high-frequency carrier signal is used in the modulation process. In this process, the low-frequency input signal changes the characteristics of the high-frequency carrier in a certain manner, depending on the modulation technique. Furthermore, as the size and speed of digital data networks continue to expand, bandwidth efficiency becomes increasingly important. This is especially true for broadband communication, where the digital signal processing is done keeping in mind the available bandwidth resources.

Hence, modulation is a very important step in the transmission of information. The information can be either analog of digital, where the carrier is a high-frequency sinusoidal waveform. As stated earlier, the input signal (analog or digital) changes the characteristics of the carrier waveform. Therefore, there are two basic modulation schemes as listed below:

- Modulation by analog signals
- Modulation by digital signals

This book presents a comprehensive overview of these modulation techniques in use today. Numerous illustrations are used to bring students up-to-date in key concepts and underlying principles of various analog and digital modulation techniques. For a head start, brief descriptions of each of these modulation techniques are presented below:

6.2   Modulation by Analog Signals

6.2.1   AM, FM, and PM

For analog signals, there are three well-known modulation techniques as listed below:

- Amplitude modulation (AM)
- Frequency modulation (FM)
- Phase modulation (PM)

By inventing the wireless transmitter or radio in 1897, the Italian physicist Tomaso Guglielmo Marconi added a new dimension to the world of communications [3, 4]. This enabled the transmission of the human voice through space without wires. For this epoch-making invention, this illustrious scientist was honored with the Nobel Prize for Physics in 1909. Even today, students of wireless or radio technology remember this distinguished physicist with reverence. A new era began in radio communications. The classical Marconi radio used a modulation technique known today as "amplitude modulation" or just AM. In AM, the amplitude of the carrier changes in accordance with the input analog signal, while the frequency of the carrier remains the same. This is shown in Fig. 6.2 where

Voltage

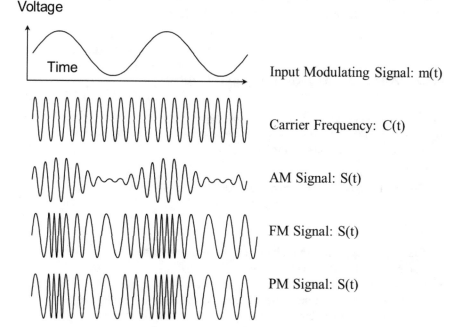

Input Modulating Signal: m(t)

Carrier Frequency: C(t)

AM Signal: S(t)

FM Signal: S(t)

PM Signal: S(t)

Fig. 6.2 Modulation by analog signal

- $m(t)$ is the input modulating audio signal
- $C(t)$ is the carrier frequency
- $S(t)$ is the AM modulated carrier frequency

As shown in the figure, the audio waveform changes the amplitude of the carrier to determine the envelope of the modulated carrier. This enables the receiver to extract the audio signal by demodulation. Notice that the amplitude of the carrier changes in accordance with the input signal, while the frequency of the carrier does not change after modulation. It can be shown that the modulated carrier $S(t)$ contains several spectral components, requiring frequency domain analysis, which will be addressed in Chap. 2. It may be noted that AM is vulnerable to signal amplitude fading.

In frequency modulation (FM), the frequency of the carrier changes in accordance with the input modulation signal as shown in Fig. 6.2 [5]. Notice that in FM, only the frequency changes, while the amplitude remains the same. Unlike AM, FM is more robust against signal amplitude fading. For this reason, FM is more attractive in commercial FM radio. In Chap. 3, it will be shown that in FM, the modulated carrier contains an infinite number of side band due to modulation. For this reason, FM is also bandwidth inefficient.

Similarly, in phase modulation (PM), the phase of the carrier changes in accordance with the phase of the carrier, while the amplitude of the carrier does not change. PM is closely related to FM. In fact, FM is derived from the rate of change of phase of the carrier frequency. Both FM and PM belong to the same mathematical family. We will discuss this more elaborately in Chap. 3.

6.2.2 AM and FM Bandwidth at a Glance

The bandwidth occupied by the modulated signal depends on bandwidth of the input signal and the modulation method as shown in Fig. 6.3. Note that the unmodulated carrier itself has zero bandwidth.

In AM:

- The modulated carrier has two side bands (upper and lower).
- Total bandwidth = 2 × base band.

In FM:

- The carrier frequency shifts back and forth from the nominal frequency by Δf, where Δf, is the frequency deviation.
- During this process, the modulated carrier creates an infinite number of spectral components, where higher order spectral components are negligible.
- The approximate FM bandwidth is given by the Carson's rule:

Fig. 6.3 Bandwidth
occupancy in AM, FM, and
PM signals

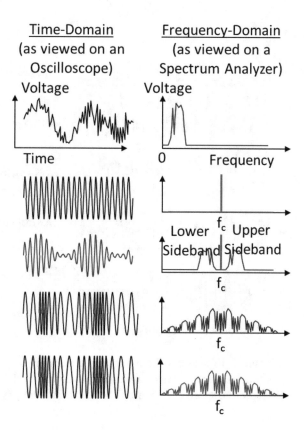

- FM BW= $2f(1+\beta)$
- f = base band frequency
- β = modulation index
- $\beta = \Delta f/f$
- Δf = frequency deviation

6.3   Modulation by Digital Signal

For digital signals, there are several modulation techniques available. The three main
digital modulation techniques are:

- Amplitude shift keying (ASK)
- Frequency shift keying (FSK)
- Phase shift keying (PSK)

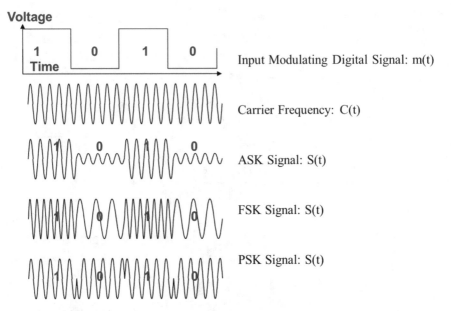

Fig. 6.4 Modulation by digital signal

Figure 6.4 illustrates the modulated waveforms for an input modulating digital signal. Brief descriptions of each of these digital modulation techniques along with the respective spectral responses and bandwidth are presented below:

6.3.1 Amplitude Shift Keying (ASK) Modulation

Amplitude shift keying (ASK), also known as on-off-keying (OOK), is a method of digital modulation that utilizes amplitude shifting of the relative amplitude of the career frequency [6–8]. The signal to be modulated and transmitted is binary; this is referred to as ASK, where the amplitude of the carrier changes in discrete levels, in accordance to the input signal, as shown

- Binary 0 (Bit 0): Amplitude = low
- Binary 1 (Bit 1): Amplitude = high

    Figure 6.4 shows the AS K modulated waveform, where

- Input digital signal is the information we want to transmit.
- Carrier is the radio frequency without modulation.
- Output is the ASK modulated carrier, which has two amplitudes corresponding to the binary input signal. For binary signal 1, the carrier is ON. For the binary signal 0, the carrier is OFF. However, a small residual signal may remain due to noise, interference, etc.

6.3.2   *Frequency Shift Keying (FSK) Modulation*

Frequency shift keying (FSK) is a method of digital modulation that utilizes frequency shifting of the relative frequency content of the signal [6–8]. The signal to be modulated and transmitted is binary; this is referred to as binary FSK (BFSK), where the carrier frequency changes in discrete levels, in accordance with the input signal as shown below:

- Binary 0 (Bit 0): Frequency $= f + \Delta f$
- Binary 1 (Bit 1): Frequency $= f - \Delta f$

   Figure 6.4 shows the FSK modulated waveform, where

- Input digital signal is the information we want to transmit.
- Carrier is the radio frequency without modulation.
- Output is the FSK modulated carrier, which has two frequencies ω_1 and ω_2, corresponding to the binary input signal.
- These frequencies correspond to the messages binary 0 and 1, respectively.

6.3.3   *Phase Shift Keying (PSK) Modulation*

Phase shift keying (PSK) is a method of digital modulation that utilizes phase of the carrier to represent digital signal [6–8]. The signal to be modulated and transmitted is binary; this is referred to as binary PSK (BPSK), where the phase of the carrier changes in discrete levels, in accordance with the input signal as shown below:

- Binary 0 (Bit 0): $Phase_1 = 0$ deg.
- Binary 1 (Bit 1): $Phase_2 = 180$ deg.

   Figure 6.4 shows the modulated waveform, where

- Input digital signal is the information we want to transmit.
- Carrier is the radio frequency without modulation.
- Output is the BPSK modulated carrier, which has two phases φ_1 and φ_2 corresponding to the two information bits.

6.4   Bandwidth Occupancy in Digital Modulation

In wireless communications, the scarcity of RF spectrum is well-known. For this reason, we have to be vigilant about using transmission bandwidth. The transmission bandwidth depends on:

- Spectral response of the encoded data
- Spectral response of the carrier frequency
- Modulation type (ASK, FSK, PSK), etc.

Let's take a closer look!

6.4.1 Spectral Response of the Encoded Data

In digital communications, data is generally referred to as a nonperiodic digital signal. It has two values:

- Binary-1 = High, Period = T
- Binary-0 = Low, Period = T

Also, data can be represented in two ways:

- Time domain representation and
- Frequency domain representation

The time domain representation (Fig. 3.12a), known as non-return-to-zero (NRZ), is given by:

$$V(t) = V \quad < 0 < t < T$$
$$= 0 \quad \text{elsewhere} \tag{6.1}$$

The frequency domain representation is given by: "Fourier transform" [9]:

$$V(\omega) = \int_0^T V.e^{-j\omega t} dt \tag{6.2}$$

$$|V(\omega)| = VT \left[\frac{\text{Sin}(\omega T/2)}{\omega T/2} \right] \tag{6.3}$$

$$P(\omega) = \left(\frac{1}{T}\right)|V(\omega)|^2 = V^2 T \left[\frac{\text{Sin}(\omega T/2)}{\omega T/2} \right]^2$$

Here, $P(\omega)$ is the power spectral density. This is plotted in Fig. 3.12b. The main lobe corresponds to the fundamental frequency, and side lobes correspond to harmonic components. The bandwidth of the power spectrum is proportional to the frequency. In practice, the side lobes are filtered out since they are relatively insignificant with respect to the main lobe. Therefore, the one-sided bandwidth is given by the ratio $f/fb = 4$. In other words, the one-sided bandwidth $= f = f_b$, where $f_b = R_b = 1/T$, T being the bit duration (Fig. 6.5).

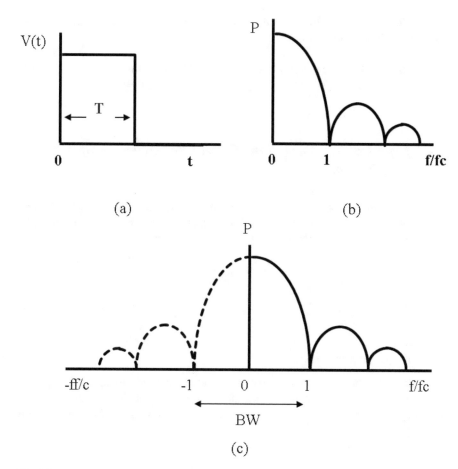

Fig. 6.5 (**a**) Discrete time digital signal (**b**) it's one-sided power spectral density and (**c**) two- sided power spectral density. The bandwidth associated with the non-return-to-zero (NRZ) data is $2R_b$, where R_b is the bit rate

The general equation for two-sided response is given by:

$$V(\omega) = \int\limits_{-\infty}^{\infty} V(t).e^{-j\omega t} dt$$

In this case, $V(\omega)$ is called two-sided spectrum of $V(t)$. This is due to both positive and negative frequencies used in the integral. The function can be a voltage or a current Fig. 3.12c shows the two-sided response, where the bandwidth is determined by the main lobe as shown below:

$$\text{Two sided bandwidth (BW)} = 2R_b (R_b = \text{Bit rate before coding}) \qquad (6.4)$$

6.4.2   Spectral Response of the Carrier Frequency Before Modulation

A carrier frequency is essentially a sinusoidal waveform, which is periodic and continuous with respect to time. It has one frequency component. For example, the sine wave is described by the following time domain equation:

$$V(t) = V_p \mathrm{Sin}(\omega t_c) \qquad (6.5)$$

where
   $Vp = \text{peak voltage}$

- $\omega_c = 2\pi f_c$
- $f_c = \text{carrier frequency in Hz}$

Figure 3.13 shows the characteristics of a sine wave and its spectral response. Since the frequency is constant, its spectral response is located in the horizontal axis, and the peak voltage is shown in the vertical axis. The corresponding bandwidth is zero (Fig. 6.6).

6.4.3   ASK Bandwidth at a Glance

In ASK, the amplitude of the carriers changes in discrete levels, in accordance with the input signal, where

Input data: $m(t) = 0 \text{ or } 1$
Carrier frequency: $C(t) = Ac\ \mathrm{Cos}(\omega_c t)$
Modulated carrier: $S(t) = m(t)C(t) = m(t)A_c\ \mathrm{Cos}(\omega_c t)$

Since $m(t)$ is the input digital signal and it contains an infinite number of harmonically related sinusoidal waveforms and that we keep the fundamental and filter out the higher-order components, we write:

$$m(t) = A_m \mathrm{Sin}(\omega_m t)$$

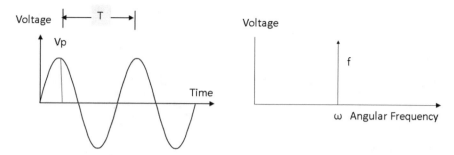

Fig. 6.6 A sine wave and its frequency response

The ASK modulated signal then becomes:

$$S(t) = m(t)S(t) = A_m A_c \mathrm{Sin}(\omega_m t)\mathrm{Cos}(\omega_c)$$
$$= A_m A_c \mathrm{Cos}\,(\omega_c \pm \omega_m)$$

The spectral response is depicted in Fig. 3.14. Notice that the spectral response after ASK modulation is the shifted version of the NRZ data. Bandwidth is given by:
BW$= 2R_b$ (coded), where R_b is the coded bit rate (Fig. 6.7).

6.4.4 FSK Bandwidth at a Glance

In FSK, the frequency of the carrier changes in two discrete levels, in accordance to the input signals. We have:

Input data: $\qquad m(t) = 0$ or 1
Carrier frequency: $\quad C(t) = A\,\mathrm{Cos}\,(\omega t)$
Modulated carrier: $\quad S(t) = A\,\mathrm{Cos}(\omega - \Delta\omega)t$, For $m(t)=1$
$\qquad\qquad\qquad\quad S(t) = A\,\mathrm{Cos}(\omega + \Delta\omega)t$, For $m(t) = 0$

where

- $S(t) =$ the modulated carrier
- $A =$ amplitude of the carrier
- $\omega =$ nominal frequency of the carrier frequency
- $\Delta\omega =$ frequency deviation

The spectral response is depicted in Fig. 3.15. Notice that the carrier frequency after FSK modulation varies back and forth from the nominal frequency f_c by $\pm\,\Delta f_c$, where Δfc is the frequency deviation. The FSK bandwidth is given by:
BW$=2(f_b + \Delta f_c) = 2f_b(1+\Delta f_c/f_b) = 2f_b(1+\beta)$, where $\beta = \Delta f/f_b$ is known as the modulation index, and f_b is the coded bit frequency (bit rate R_b) (Fig. 6.8).

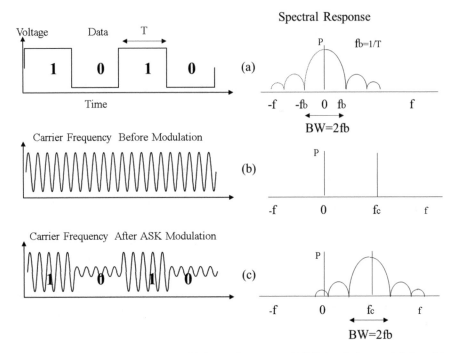

Fig. 6.7 ASK bandwidth at a glance. (**a**) Spectral response of NRZ data before modulation. (**b**) Spectral response of the carrier before modulation. (**c**) Spectral response of the carrier after modulation. The transmission bandwidth is 2*fb*, where *fb* is the bit rate and *T*=1/*fb* is the bit duration for NRZ data

6.4.5   BPSK Bandwidth at a Glance

In BPSK, the phase of the carrier changes in two discrete levels, in accordance to the input signal. Here we have:

Input data: $m(t) = 0$ or 1
Carrier frequency: $C(t) = A \cos(\omega t)$
Modulated carrier: $S(t) = A \cos(\omega + \varphi)t$

where

- A = amplitude of the carrier frequency
- ω = angular frequency of the carrier
- φ) = phase of the carrier frequency

The table below shows the number of phases and the corresponding bits per phase for MPSK modulation schemes for M=2, 4, 8, 16, 32, 64, etc. It will be shown that higher-order MPSK modulation schemes (M>2) are spectrally efficient. See Problem 3.7.

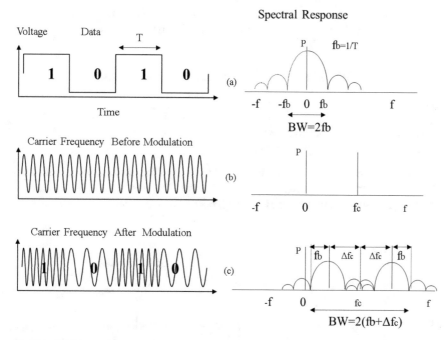

Fig. 6.8 FSK bandwidth at a glance. (**a**) Spectral response of NRZ data before modulation. (**b**) Spectral response of the carrier before modulation. (**c**) Spectral response of the carrier after modulation. The transmission bandwidth is $2(f_b + \Delta f_c)$. f_b is the bit rate, and Δf_c is the frequency deviation $= 1/f_b$ is the bit duration for NRZ data

Modulation	Number of phases φ	Number of bits per phase
BPSK	2	1
QPSK	4	2
8PSK	8	3
16	16	4
32	32	5
64	64	6
:	:	:

Figure 3.16 shows the spectral response of the BPSK modulator. Since there are two phases, the carrier frequency changes in two discrete levels, one bit per phase, as follows:

$$\varphi = 0° \text{ for bit } 0$$

$$\varphi = 180° \text{ for bit } 1$$

Notice that the spectral response after BPSK modulation is the shifted version of the NRZ data, centered on the carrier frequency f_c. The transmission bandwidth is given by:

$$\mathrm{BW(BPSK)} = 2R_b/\mathrm{Bit\ per\ Phase} = 2R_b/1 = 2R_b$$

where

- R_b is the coded bit rate (bit frequency)
- For BPSK, $\varphi = 2$, one bit per phase

Also, notice that the BPSK bandwidth is the same as the one in ASK modulation. This is due to the fact that the phase of the carrier changes in two discrete levels, while the frequency remains the same (Fig. 6.9).

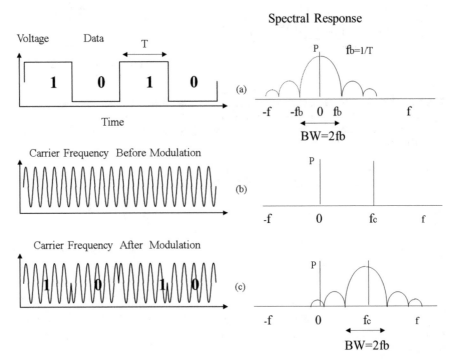

Fig. 6.9 PSK bandwidth at a glance. (**a**) Spectral response of NRZ data before modulation. (**b**) Spectral response of the carrier before modulation. (**c**) Spectral response of the carrier after modulation

6.5 Conclusions

This chapter presents a brief overview of modulation techniques covered in this book. Numerous illustrations are used to bring students up-to-date in key concepts and underlying principles of various analog and digital modulation techniques. In particular, the following topics will be presented in this book:

- Amplitude Modulation (AM)
- Frequency Modulation (FM)
- Bandwidth Occupancy in AM and FM
- Amplitude Shift Keying (ASK)
- Frequency Shift Keying (FSK)
- Phase Shift Keying (PSK)
- Bandwidth Occupancy in ASK, FSK, and PSK

References

1. S. Faruque, *Radio Frequency Source Coding Made Easy* (Springer, Cham, 2014) ISBN: xxxxxxx
2. S. Faruque, *Radio Frequency Channel Coding Made Easy* (Springer, Cham, 2014) ISBN: xxxxxxx
3. G. Marconi, British patent No. 12,039. *Improvements in Transmitting Electrical impulses and Signals, and in Apparatus therefor.* Date of Application 2 June 1896; Complete Specification Left, 2 March 1897; Accepted, 2 July 1897 (later claimed by Oliver Lodge to contain his own ideas which he failed to patent), 1897
4. G. Marconi, British patent No. 7,777. *Improvements in Apparatus for Wireless Telegraphy.* Date of Application 26 April 1900; Complete Specification Left, 25 February 1901; Accepted, 13 April 1904, 1900
5. E.H. Armstrong, A method of reducing disturbances in radio signaling by a system of frequency modulation. Proc. IRE (IRE) **24**(5), 689–740 (May 1936). https://doi.org/10.1109/JRPROC.1936.227383
6. D.R. Smith, *Digital Transmission Systems* (Van Nostrand Reinhold Co, New York, 1985) ISBN: 0442009178
7. W. Leon, I.I. Couch, *Digital and Analog Communication Systems*, 7th edn. (Prentice-Hall Inc., Englewood Cliffs, 2004) ISBN: 0-13-142492-0
8. B. Sklar, *Digital Communications Fundamentals and Applications* (Prentice Hall, Upper Saddle River, 1988)
9. J.B.J. Fourier, A. Freeman, translator, *The Analytical Theory of Heat* (The University Press, 1878)

Chapter 7
Amplitude Modulation (AM)

Topics
- Introduction
- Amplitude Modulation (AM)
- AM Spectrum and Bandwidth
- Double Side Band Suppressed Carrier (DSBSC)
- DSBSC Spectrum and Bandwidth
- Single Side Band (SSB) Carrier
- SSB Spectrum and Bandwidth

7.1 Introduction

By inventing the wireless transmitter or radio in 1897, the Italian physicist Tomaso Guglielmo Marconi added a new dimension to the world of communications [1, 2]. This enabled the transmission of the human voice through space without wires. For this epoch-making invention, this illustrious scientist was honored with the Nobel Prize for Physics in 1909. Even today, students of wireless or radio technology remember this distinguished physicist with reverence. A new era began in radio communications.

The classical Marconi radio used a modulation technique known today as "amplitude modulation" or just AM, which is the main topic in this chapter. In AM, the amplitude of the carrier changes in accordance with the input analog signal, while the frequency of the carrier remains the same. This is shown in Fig. 7.1 where:

- $m(t)$ is the input modulating audio signal
- $C(t)$ is the carrier frequency
- $S(t)$ is the AM modulated carrier frequency

Fig. 7.1 AM waveforms.
The amplitude of the carrier
changes in accordance with
the input analog signal. The
frequency of the carrier
remains the same

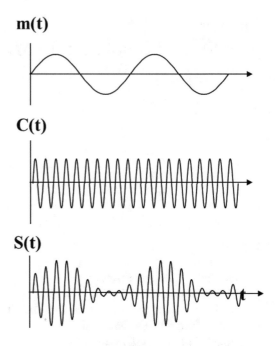

As shown in the figure, the audio waveform changes the amplitude of the carrier to determine the envelope of the modulated carrier. This enables the receiver to extract the audio signal by demodulation. Notice that the amplitude of the carrier changes in accordance with the input signal, while the frequency of the carrier does not change after modulation. However, it can be shown that the modulated carrier S (t) contains several spectral components, requiring frequency domain analysis. In an effort to examine this, this chapter will present the following topics:

- Amplitude modulation (AM) and AM spectrum
- Double side band suppressed carrier (DSBSC) and DSBSC spectrum
- Single side band (SSB) carrier and SSB spectrum

In the following sections, the above disciplines in AM modulation will be presented along with their respective spectrum and bandwidth. These materials have been augmented by diagrams and associated waveforms to make them easier for readers to grasp.

7.2   Amplitude Modulation

Amplitude modulation (AM) is a method of analog modulation that utilizes amplitude variations of the relative amplitude of the career frequency [3–5]. The signal to be modulated and transmitted is analog. This is referred to as AM, where the amplitude of the carrier changes in accordance with the input signal.

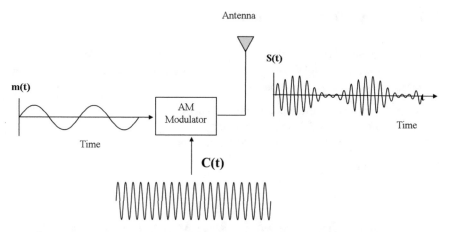

Fig. 7.2 Illustration of amplitude modulation. The amplitude of the carrier $C(t)$ changes in accordance with the input modulating signal $m(t)$. $S(t)$ is the modulated waveform which is transmitted by the antenna

Figure 7.2 shows a functional diagram of a typical AM modulator for a single tone. Here, $m(t)$ is the input analog signal we want to transmit, $C(t)$ is the carrier frequency without modulation, and $S(t)$ is the output AM modulated carrier frequency. These parameters are described below:

$$\begin{aligned} m(t) &= A_{\mathrm{m}}\mathrm{Cos}(2\pi f_{\mathrm{m}}t) \\ C(t) &= A_{\mathrm{c}}\mathrm{Cos}(2\pi f_{c}t) \; f_{\mathrm{c}} \gg f_{\mathrm{m}} \\ S(t) &= [1 + m(t)]\, C(t) \\ &= C(t) + m(t)\, C(t) \end{aligned} \tag{7.1}$$

Therefore,
When $m(t) = 0$:

$$S(t) = A_{\mathrm{c}}\mathrm{Cos}(2\pi f_{c}t) \tag{7.2}$$

When $m(t) = A_{\mathrm{m}}\, \mathrm{Cos}(2\pi f_{\mathrm{m}}t)$:

$$S(t) = A_{\mathrm{c}}\mathrm{Cos}\,(2\pi f_{c}t) + A_{\mathrm{c}}A_{\mathrm{m}}\mathrm{Cos}\,(2\pi f_{\mathrm{m}}t)\;\mathrm{Cos}\,(2\pi f_{c}t) \tag{7.3}$$

In the above equation, we see that:

- The first term is the carrier only, which does not have information
- The 2nd term contains the information, which has several spectral components, requiring further analysis to quantify them

7.3   AM Spectrum and Bandwidth

In wireless communications, the scarcity of RF spectrum is well-known. For this reason, we have to be vigilant about using transmission bandwidth and modulation. The transmission bandwidth depends on:

- Spectral response of the input modulating signal
- Spectral response of the carrier frequency
- Modulation type (AM, FM ASK, FSK, PSK, etc.)

    Let's take a closer look!

7.3.1   Spectral Response of the Input Modulating Signal

In AM, the input modulating signal is a continuous time low-frequency analog signal. For simplicity, we use a sinusoidal waveform, which is periodic and continuous with respect to time. It has one frequency component. For example, the sine wave is described by the following time domain equation:

$$V(t) = V_{\mathrm{p}}\mathrm{Sin}(\omega_{\mathrm{m}}t) \tag{7.4}$$

where

- $V\mathrm{p} = $ peak voltage
- $\omega_{\mathrm{m}} = 2\pi f_{\mathrm{m}}$
- $f_{\mathrm{m}} = $ input modulating frequency in Hz

    Figure 7.3 shows the characteristics of a sine wave and its spectral response. Since the frequency is constant, its spectral response is located in the horizontal axis at f_{m}, and the peak voltage is shown in the vertical axis. The corresponding bandwidth is zero.

7.3.2   Spectral Response of the Carrier Frequency

A carrier frequency (f_{c}) is essentially a sinusoidal waveform, which is periodic and continuous with respect to time. It has one frequency component, which is much higher than the input modulating frequency ($f_{\mathrm{c}} \gg f_{\mathrm{m}}$). For example, the sine wave is described by the following time domain equation:

$$V(t) = V_{\mathrm{p}}\mathrm{Sin}(\omega_{\mathrm{c}}t) \tag{7.5}$$

where

m(t)

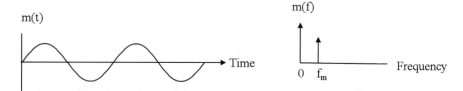

Fig. 7.3 A low-frequency sine wave and its frequency response

C(t)

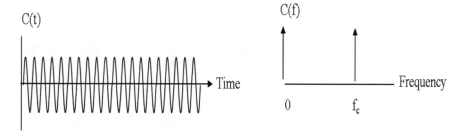

Fig. 7.4 A high-frequency sine wave and its frequency response

Vp = peak voltage

- $\omega_c = 2\pi f_c$
- f_c = carrier frequency in Hz

Figure 7.4 shows the characteristics of a sine wave and its spectral response. Since the frequency is constant, its spectral response is located in the horizontal axis at f_c and the peak voltage is shown in the vertical axis. The corresponding bandwidth is zero.

7.3.3   AM Spectrum and Bandwidth

Let's consider the AM signal again, which was derived earlier:

$$S(t) = A_c \text{Cos}\,(2\pi f_c t) + A_c A_m \text{Cos}(2\pi f_m t)\,\text{Cos}\,(2\pi f_c t) \qquad (7.6)$$

Using the following trigonometric identity:

$$\text{CosA CosB} = 1/2\text{Cos}\,(A + B) + 1/2\text{Cos}\,(A - B) \qquad (7.7)$$

where

- $A = 2\pi\,(f_c + f_m)t$
- $B = 2\pi\,(f_c - f_m)t$

we get

$$S(t) = A_c\mathrm{Cos}\,(2\pi f_c t) + (1/2)A_c A_m \mathrm{Cos}\,[2\pi(\,f_c + f_m)t] \\ + (1/2)A_c A_m \mathrm{Cos}[2\pi(\,f_c - f_m)t] \tag{7.8}$$

This is the spectral response of the AM modulated signal. It has three spectral components:

- The carrier: fc
- Upper side band: $f_c + f_m$
- Lower side band: $f_c - f_m$

where f_c is the carrier frequency and f_m is the input modulating frequency. This is shown in Fig. 7.5. The AM bandwidth (BW) is given by

$$\mathrm{BW} = 2f_m \tag{7.9}$$

Notice that the power is distributed among the side bands and the carrier, where the carrier does not contain any information. Only the side bands contain the information. Therefore, AM is inefficient in power usage.

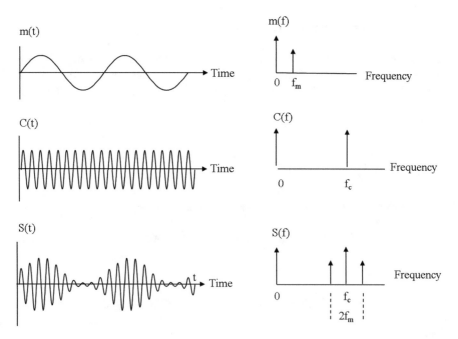

Fig. 7.5 AM spectrum. The bandwidth is given by 2 fm

7.3.4 AM Response due to Low and High Modulating Signals

If $m(t)$ has a peak positive value of less than $+1$ and peak negative value of higher than -1, the modulation is less than 100%. This is shown in Fig. 7.7.

On the other hand, if $m(t)$ has a peak positive value of $+1$ and peak negative value of -1, the modulation is 100% [3–5]. Therefore,

$$\text{For } m(t) = -1 : S(t) = A_c[1 - 1]\,\text{Cos}\,(2\pi f_c t) = 0 \tag{7.10}$$

$$\begin{aligned} \text{For } m(t) = +1 : S(t) &= A_c[1 + 1]\,\text{Cos}\,(2\pi f_c t)\\ &= 2A_c\text{Cos}\,(2\pi f_c t) \end{aligned} \tag{7.11}$$

This is called 100% modulation, as shown in Fig. 7.6. The percent modulation is described by the following equation:

The overall modulation percentage is :

$$\begin{aligned} \%\ \textit{Overall Modulation} &= \frac{A_{\max} - A_{\min}}{A_c} \times 100\\ &= \frac{\max\,[m(t)] - \min\,[m(t)]}{2A_c} \times 100 \end{aligned} \tag{7.12}$$

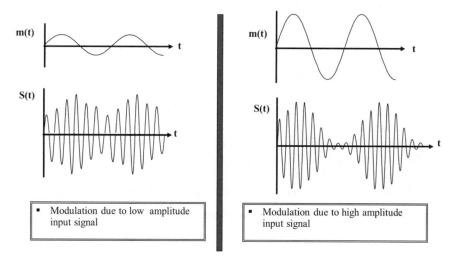

Fig. 7.6 Amplitude modulation due to low and high modulating signals

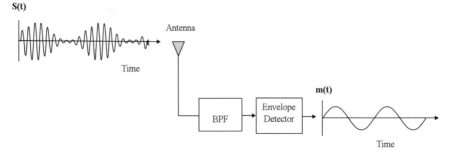

Fig. 7.7 AM demodulation technique. As the signal enters the receiver, it passes through the bandpass filter, which is tuned to the carrier frequency f_o. Next, the recovered signal is passed through an envelope detector to recover the original signal that was transmitted

7.3.5   AM Demodulation

Once the modulated analog signal has been transmitted, it needs to be received and demodulated. This is accomplished by the use of a bandpass filter that is tuned to the appropriate carrier frequency. Figure 7.7 shows the conceptual model of the AM receiver. As the signal enters the receiver, it passes through the bandpass filter, which is tuned to the carrier frequency f_o. Next, the recovered signal is passed through an envelope detector to recover the original signal that was transmitted.

7.3.6   Drawbacks in AM

- The modulated signal contains the carrier; carrier takes power and it does not have the information.
- Therefore, AM is inefficient in power usage.
- Moreover, there are two side bands, containing the same information.
- It is bandwidth inefficient.
- AM is also susceptible to interference, since it affects the amplitude of the carrier.

Therefore, a solution is needed to improve bandwidth and power efficiency.

Problem 7.1

Given:

- Input modulating frequency fm = 10 kHz
- Carrier frequency fc = 400 kHz

Find:

- Spectral components
- Bandwidth

Solution:

Spectral components:

- fc = 400 kHz
- fc + fm = 400 kHz + 10 kHz = 410 kHz
- fc − fm = 400 kHz − 10 kHz = 390 kHz

Bandwidth:

- BW = 2 fm = 2 × 10 kHz = 20 kHz

7.4   Double Side Band Suppressed Carrier (DSBSC)

7.4.1   DSBSC Modulation

Double side band suppressed carrier (DSBSC), also known as product modulator, is an AM signal that has a suppressed carrier [3–5]. Let's take the original AM signal once again, as given below:

$$
\begin{aligned}
S(t) = & A_c \cos\ (2\pi f_c t) + (1/2)A_c A_m \cos\ [2\pi(\ f_c + f_m)t] \\
& + \ (1/2)A_c A_m \cos\ [2\pi(\ f_c - f_m)t]
\end{aligned}
\tag{7.13}
$$

Notice that there are three spectral components:

- The first term is the carrier only, which does not have any information.
- The second and third terms contain information.

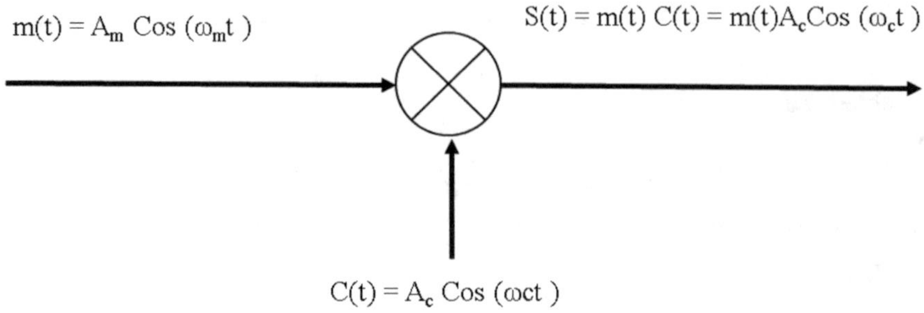

$m(t) = A_m \, \text{Cos} \, (\omega_m t)$ $S(t) = m(t) \, C(t) = m(t)A_c\text{Cos} \, (\omega_c t)$

$C(t) = A_c \, \text{Cos} \, (\omega ct)$

Fig. 7.8 Symbolic representation of DSBSC, also known as product modulator

In DSBSC, we suppress the carrier, which is the first term that does not have any information. Therefore, by suppressing the first term, we obtain

$$S(t) = (1/2)A_cA_m \cos \left[2\pi(f_c + f_m)t\right] + (1/2)A_cA_m \cos \left[2\pi(f_c - f_m)t\right] \quad (7.14)$$

Next, we use the following trigonometric identities:

- $\text{Cos}(A + B) = \text{CosA CosB} - \text{SinA SinB}$
- $\text{Cos}(A - B) = \text{CosA CosB} + \text{SinA SinB}$

  With $A = 2\pi f_m t = \omega_m t$ and $B = 2\pi f_c t = \omega_c t$, we obtain

$$S(t) = A_cA_m\text{Cos} \, (\omega_m t) \, \text{Cos} \, (\omega_c t) \quad (7.15)$$

Now, define:

- $m(t) = A_m \, \text{Cos} \, (\omega_m t)$
- $C(t) = A_c \, \text{Cos} \, (\omega_c t)$

  Then, we can write the above equations as

$$S(t) = m(t) \, C(t) \quad (7.16)$$

This is the DSBSC waveform. Since the output is the product of two signals, it is also known as product modulator. The symbolic representation is given in Fig. 7.8, where $m(t)$ is the input modulating signal and $C(t)$ is the carrier frequency.

7.4.2   Generation of DSBSC Signal

A DSBSC signal can be generated by using two AM modulators arranged in a balanced configuration as shown in Fig. 7.9 [3–5]. The outcome is a cancellation of

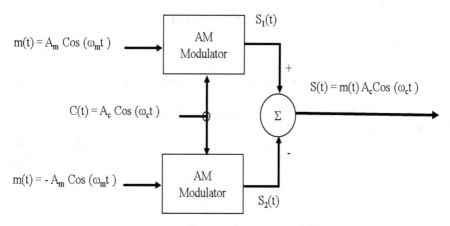

Fig. 7.9 Construction of DSBSC modulator. The output is the product of two signals

the discrete carrier. Also, the output is the product of two inputs: $S(t) = m(t)\, C(t)$. This is why it is called "product modulator."

Proof of DSBSC

Consider the DSBSC modulator as shown in Fig. 7.9. Here, the AM modulators generate $S_1(t)$ and $S_2(t)$, which are given by

$$S_1(t) = A_c[1 + m(t)]\, \text{Cos}\,(\omega_c t)$$
$$S_2(t) = A_c[1 - m(t)]\, \text{Cos}\,(\omega_c t) \tag{7.17}$$

Subtracting $S_2(t)$ from $S_1(t)$, we essentially cancel the carrier to obtain

$$S(t) = S_1(t) - S_2(t)$$
$$= 2m(t)\, A_c \text{Cos}\,(\omega_c t) \tag{7.18}$$

Therefore, except for the scaling factor 2, the above equation is exactly the same as the desired DSBSC waveform shown earlier, which does not have the carrier. In other words, the carrier has been suppressed, hence the name double side band suppressed carrier (DSBSC).

7.4.3   DSBSC Spectrum and Bandwidth

We begin with the DSBSC modulated signal:

$$S(t) = 2\ m(t) A_c \text{Cos}\,(\omega_c t) \tag{7.19}$$

where

$$m(t) = A_m Cos\ (\omega_m t)$$

Therefore,

$$S(t) = 2A_c A_m Cos\ (\omega_m t)\ Cos\ (\omega_c t) \tag{7.20}$$

This is the desired DSBSC waveform for spectral analysis.
Now, use the trigonometric identity:

$$Cos A\ Cos B = \tfrac{1}{2} Cos\ (A + B) + \tfrac{1}{2} Cos\ (A - B) \tag{7.21}$$

where

$$A = \omega_m t = 2\pi f_m t \text{ and } B = \omega_c t = 2\pi f_c t$$

Therefore,

$$\begin{aligned} S(t) &= A_c[\ Cos\ (\omega_c + \omega_m)t + Cos\ (\omega_c - \omega_m)t] \\ &= A_c[\ Cos\ 2\pi(\ f_c + f_m)t + Cos\ 2\pi(\ f_c - f_m)t] \end{aligned} \tag{7.22}$$

Notice that the carrier power is distributed among the side bands (Fig. 7.10). Therefore, it is more efficient. The bandwidth is given by

$$BW = 2fm \tag{7.23}$$

7.4.4 DSBSC Drawback

- There are two identical side bands.
- Each side band contains the same information.
- Bandwidth is $2f_m$.
- Unnecessary power usage.

Therefore, a solution is needed to improve bandwidth efficiency.

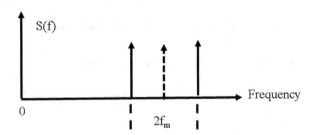

Fig. 7.10 DSBSC spectrum, where the carrier frequency is suppressed. The bandwidth is given by 2 fm

Problem 7.2

Given:

Two product modulators using identical carriers are connected in a series, as shown below:

$m(t) = A_m \, Cos(\omega_m t)$ — $S_1(t)$ — $S_2(t)$

$C(t) = A_c \, Cos \, (\omega_c t)$ $C(t) = A_c \, Cos \, (\omega_c t)$

Find:

(a) The output waveform $S_2(t)$
(b) What is the function of this circuit?

Solution:

(a)
$$S1(t) = m(t)C(t)$$
$$S2(t) = S1(t)C(t) = m(t)C^2(t) = A_m Cos(\omega_m t) \, Ac^2 Cos2(\omega_c t)$$
(b) The function of the circuit is to demodulate DSBSC signals, where the carrier frequency is filtered out.

7.5   Single Side Band (SSB) Modulation

7.5.1   Why SSB Modulation?

- The basic AM has a carrier which does not carry information—*inefficient power usage.*
- The basic AM has two side bands contain the same information—*additional loss of power.*
- DSBSC has two side bands, containing the same information—*loss of power.*
- Therefore, the basic AM and DSBSC are bandwidth and power inefficient.
- SSB is bandwidth and power efficient.

7.5.2   Generation of SSB Modulated Signal

Single side band (SSB) modulation uses two product modulators as shown in Fig. 7.11 [3–5], where.

$$m(t) = A_m \text{Cos} (\omega_m t) \tag{7.24}$$

$$m(t)* = A_m \text{Sin} (\omega_m t) - (\text{Hilbert Transform}) \tag{7.25}$$

$$C(t) = A_c \text{Cos} (\omega_c t) \tag{7.26}$$

$$C(t)* = A_c \text{Sin} (\omega_c t) - (\text{Hilbert Transform}) \tag{7.27}$$

Solving for S1, S2, and S3, we obtain:

$$S1(t) = A_c A_m \text{Cos} (\omega_m t) \ \text{Cos} (\omega_c t) \tag{7.28}$$

$$S2(t) =_c A_m \text{Sin} (\omega_m t) \ \text{Sin} (\omega_c t) \tag{7.29}$$

$$\begin{aligned} S3(t) &= S1(t) - S2(t) \\ &= A_c A_m \text{Cos} (\omega_m t) \ \text{Cos} (\omega_c t) - A_c A_m \ \text{Sin} (\omega_m t) \ \text{Sin} (\omega_c t) \end{aligned} \tag{7.30}$$

Using the following formula:

- Cos A Cos B = 1/2Cos (A + B) + 1/2 Cos (A − B)
- Sin A Sin B = 1/2 Cos (A − B) − 1/2 Cos (A + B)

Solving for S3, we get

$$S3 = A_c A_m \ \text{Cos} (\omega_c + \omega_m)t \tag{7.31}$$

In the above equation, S3(t) is the desired SSB signal, which is the upper side band only.

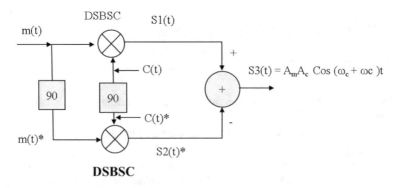

Fig. 7.11 Generation of SSB signal

Fig. 7.12 SSB spectrum showing the upper side band. The SSB bandwidth is f_m

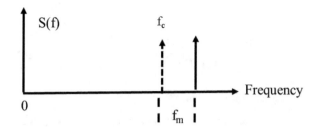

7.5.3 SSB Spectrum and Bandwidth

Let's consider the SSB signal again:

$$S3 = A_c A_m \ \text{Cos} \ (\omega_c + \omega_m)t$$
$$= A_c A_m \ \text{Cos} \ 2\pi \ (f_c + f_m)t \tag{7.32}$$

Here, we see that the SSB spectrum contains only one side band. Therefore, it is more efficient. The SSB bandwidth is given by

$$\text{SSB BW} = f_m \tag{7.33}$$

Figure 7.12 displays the SSB spectrum.

Problem 7.3
Given:

- $m(t) = A_m \ \text{Cos} \ (\omega_m t)$.
- $m(t)^* = A_m \ \text{Sin} \ (\omega_m t)$ — (Hilbert transform)
- $C(t) = A_c \ \text{Cos} \ (\omega_c t)$.
- $C(t)^* = A_c \ \text{Sin} \ (\omega_c t)$ — (Hilbert transform)

    Design an SSB modulator to realize the lower side band. Sketch the spectral response.

Solution:
Solving for S1 and S2, we obtain:

- $S1(t) = A_c \ A_m \ \text{Cos} \ (\omega_m t) \ \text{Cos} \ (\omega_c t)$
- $S2(t) = A_c \ A_m \ \text{Sin} \ (\omega_m t) \ \text{Sin} \ (\omega_c t)$

Obtain $S3$ as

$$S3(t) = S1(t) + S2(t)$$
$$= A_c A_m \text{Cos } (\omega_m t) \text{ Cos } (\omega_c t) + A_c A_m \text{ Sin } (\omega_m t) \text{ Sin } (\omega_c t)$$

Using the following formula:

- Cos A Cos B = 1/2Cos(A + B) + 1/2Cos(A − B)
- Sin A Sin B = 1/2Cos(A − B) − 1/2Cos(A + B)

Solving for $S3$, we get

$$S3 = A_c A_m \text{ Cos } (\omega_c − \omega_m)t$$
$$= A_c A_m \text{ Cos } 2\pi \ (f_c − f_m)t$$

In the above equation, $S3(t)$ is the desired SSB signal, which is the lower side band only. The spectral response, showing the lower side band, is presented below.

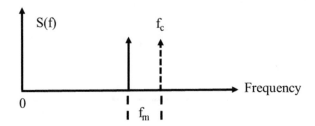

7.6   Conclusions

This chapter presents the key concepts and underlying principles of amplitude modulation. It was shown how the audio waveform changes the amplitude of the carrier to determine the envelope of the modulated carrier. It was also shown that the modulated carrier contains several spectral components that lead to DSBSC and SSB modulation techniques. In particular, the following topics were presented in this chapter:

- Amplitude Modulation (AM)
- AM spectrum and Bandwidth

- Double Side Band Suppressed Carrier (DSBSC)
- DSBSC spectrum and Bandwidth
- Single Side Band (SSB)
- SSB spectrum and Bandwidth

These materials have been augmented by diagrams and associated waveforms to make them easier for readers to grasp.

References

1. Guglielmo Marconi, British patent No. 12,039 (1897) "Improvements in Transmitting Electrical impulses and Signals, and in Apparatus therefor". Date of Application 2 June 1896; Complete Specification Left, 2 March 1897; Accepted, 2 July 1897 (later claimed by Oliver Lodge to contain his own ideas which he failed to patent)
2. Guglielmo Marconi, British patent No. 7,777 (1900) "Improvements in Apparatus for Wireless Telegraphy". Date of Application 26 April 1900; Complete Specification Left, 25 February 1901; Accepted, 13 April 1901
3. W. Leon, I.I. Couch, *Digital and Analog Communication Systems*, 7th edn. (Prentice-Hall, Inc., Englewood Cliffs, 2001). ISBN: 0-13-142492-0
4. A.P. Godse, U.A. Bakshi, *Communication Engineering* (Technical Publications, 2009), p. 36. ISBN 978-81-8431-089-4
5. W. Silver (ed.), Chapter: 14 Transceivers, in *The ARRL Handbook for Radio Communications*, 88th edn., (American Radio Relay League, 2011). ISBN 978-0-87259-096-0

Chapter 8
Frequency Modulation (FM)

Topics
- Introduction
- Frequency Modulation (FM)
- FM Spectrum
- Carson's Rule and FM Bandwidth
- Bessel Function and FM Bandwidth
- FM Bandwidth Dilemma

8.1 Introduction

In frequency modulation (FM), the frequency of the carrier changes in accordance with the input analog signal, while the amplitude of the carrier remains the same [1–5]. This is shown in Fig. 8.1 where:

- $m(t)$ is the input modulating audio signal
- $C(t)$ is the carrier frequency
- $S(t)$ is the FM modulated carrier frequency

As shown in the figure, the audio waveform changes the frequency of the carrier. This enables the receiver to extract the audio signal by demodulation. Notice that the frequency of the carrier changes in accordance with the input signal, while the amplitude of the carrier does not change after modulation. However, it can be shown that the modulated carrier $S(t)$ contains an infinite number of spectral components, requiring frequency domain analysis [3]. In an effort to examine this, this chapter will present the following topics:

- The Basic Frequency Modulation (FM)
- FM Spectrum
- FM Bandwidth

© The Editor(s) (if applicable) and The Author(s), under exclusive license to
Springer Nature Switzerland AG 2021
S. Faruque, *Free Space Laser Communication with Ambient Light Compensation*,
https://doi.org/10.1007/978-3-030-57484-0_8

Fig. 8.1 Illustration of FM

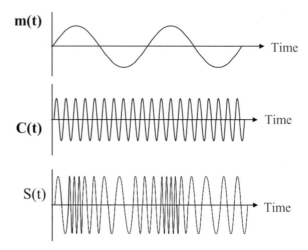

8.2   Frequency Modulation (FM)

8.2.1   *Background*

FM is a form of angle modulation, where the frequency of the carrier varies in accordance with the input signal. Here, the angle refers to the angular frequency (ω). The angular frequency ω is also recognized as angular speed or circular frequency. It's a measure of rotation rate or the rate of change of the phase of a sinusoidal waveform as illustrated in Fig. 8.2.

Magnitude of the angular frequency ω is defined by one revolution or 2π radians:

$$\omega = 2\pi f = d\theta/dt \quad \text{Radians per second} \tag{8.1}$$

where

- ω = angular frequency in radians per seconds
- f = frequency in hertz (Hz) or cycles per second
- θ = phase angle

Notice that in Eq. 8.1, the angular frequency ω is greater than the frequency f by a factor of 2π. Now solving for the phase angle, we obtain

$$\theta_i(t) = 2\pi \int_0^t f_i dt \tag{8.2}$$

- θ_i = instantaneous phase angle
- f_i = instantaneous frequency

This forms the basis of our derivation of FM as presented in the following section.

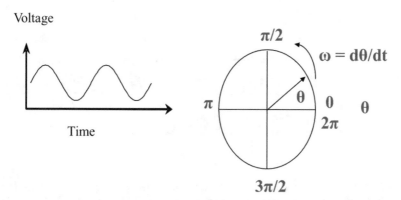

Fig. 8.2 A sinusoidal waveform in the time domain and its representation in the phase domain

8.2.2 The Basic FM

Frequency modulation (FM) is a method of analog modulation that utilizes frequency variation of the relative frequency of the career [1]. The signal to be modulated and transmitted is analog. This is referred to as FM, where the frequency of the carrier changes in accordance with the input signal. The modulated carrier frequency f_c varies back and forth and depends on amplitude A_m and frequency f_m of the input signal.

Figure 8.3 shows the functional diagram of a typical FM, using a single-tone input modulating signal. Here, $m(t)$ is the input analog signal we want to transmit, $C(t)$ is the carrier frequency without modulation, and $S(t)$ is the output FM modulated carrier frequency. These parameters are described below:

We examine this by means of a single-tone input modulating signal and its angular frequency as shown in Fig. 8.2. Since the frequency of the carrier varies in accordance with the input signal, the instantaneous frequency of the carrier is given by

$$f_i(t) = f_c(t) + D_f m(t) \qquad (8.3)$$

where

- f_i = instantaneous frequency
- f_c = carrier frequency
- D_f = constant
- $m(t) = A_m \cos(wmt)$

The FM modulated signal is given by

$$S(t) = A_c \cos(\theta_i) \qquad (8.4)$$

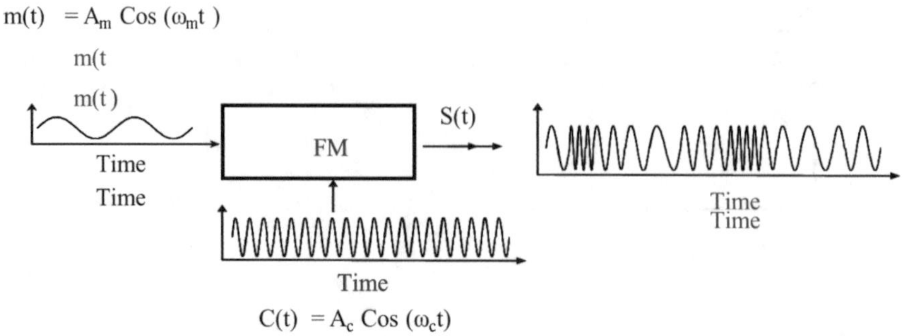

$m(t) = A_m \, Cos \, (\omega_m t \,)$

Fig. 8.3   A functional diagram of a typical FM modulator using a single-tone input modulating signal. Here, $m(t)$ is the input analog signal we want to transmit, $C(t)$ is the carrier frequency without modulation, and $S(t)$ is the output FM modulated carrier frequency

where A_c is the amplitude of the carrier frequency and θ_i is the instantaneous angle. Substituting Eq. (8.2) into Eq. (8.4) for θ_i, we get

$$S(t) = A_c Cos \left[2\pi \int_0^t f_i dt \right] \tag{8.5}$$

where f_i is the instantaneous frequency. Substituting Eq. (8.3) into Eq. (8.5) for f_i, we obtain

$$S(t) = A_c Cos \left[2\pi \int_0^t \{fc(t) + Df \, m(t) \}dt \right] \tag{8.6}$$

where $m(t) = A_m \, Cos \, (\omega_m t)$ is the input modulating signal. Integrating the above equation, we obtain the desired FM signal as follows:

$$\begin{aligned} S(t) &= A_c Cos \left[2\pi f_c t + \left(\frac{\Delta f}{f_m} \right) Sin(2\pi f_m t) \right] \\ &= A_c Cos[2\pi f_c t + \beta Sin(2\pi f_m t)] \end{aligned} \tag{8.7}$$

$$\beta = \left(\frac{\Delta f}{f_m} \right) = \text{Modulation Index} \quad \Delta f = D_f A_m = \text{Freq.Deviation}$$

where

- $S(t) =$ FM modulated carrier signal
- $f_c =$ frequency of the carrier
- $A_c =$ amplitude of the carrier frequency
- $\Delta f = D_f A_m =$ frequency deviation

- f_m = input modulating frequency
- A_m = amplitude of the input modulating signal
- D_f = A constant parameter
- $\beta = \Delta f / f_m$ = modulation index

Note that the modulation index β is an important design parameter in FM. It is directly related to FM bandwidth. It may also be noted that FM bandwidth depends on both frequency and amplitude of the input modulating signal. Let's take a closer look.

8.3 FM Spectrum and Bandwidth

In wireless communications, the scarcity of RF spectrum is well-known. For this reason, we have to be vigilant about using transmission bandwidth and modulation. The transmission bandwidth depends on:

- Spectral response of the input modulating signal
- Spectral response of the carrier frequency
- Modulation type

Let's take a closer look!

8.3.1 Spectral Response of the Input Modulating Signal

In FM, the input modulating signal is a continuous time low-frequency analog signal. For simplicity, we use a sinusoidal waveform, which is periodic and continuous with respect to time. It has one frequency component. For example, the sine wave is described by the following time domain equation:

$$V(t) = V_p \mathrm{Sin}(\omega_c t) \tag{8.8}$$

where

- V_p = peak voltage
- $\omega_m = 2\pi f_m$
- f_m = input modulating frequency in Hz

Figure 8.4 shows the characteristics of a sine wave and its spectral response. Since the frequency is constant, its spectral response is located in the horizontal axis at f_m, and the peak voltage is shown in the vertical axis. The corresponding bandwidth is zero.

m(t)

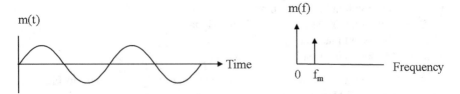

Fig. 8.4 A low-frequency sine wave and its frequency response

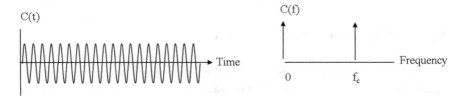

Fig. 8.5 A high-frequency sine wave and its frequency response before modulation

8.3.2   Spectral Response of the Carrier Frequency

A carrier frequency (f_c) is essentially a sinusoidal waveform, which is periodic and continuous with respect to time. It has one frequency component, which is much higher than the input modulating frequency ($f_c \gg f_m$). For example, the sine wave is described by the following time domain equation:

$$V(t) = V_p Sin(\omega_m t) \tag{8.9}$$

where

Vp = peak voltage

- $\omega_c = 2\pi f_c$
- f_c = carrier frequency in Hz

Figure 8.5 shows the characteristics of a high-frequency sine wave and its spectral response. Since the frequency is constant, its spectral response is located in the horizontal axis at f_c, and the peak voltage is shown in the vertical axis. The corresponding bandwidth is zero.

8.3.3 FM Spectrum

In FM, the frequency of the carrier changes in accordance with the input signal. Here, we have:

$$\text{Input Signal :} \quad m(t) = A_m \text{Cos}\,(\omega_m t)$$
$$\text{Carrier Frequency :} \quad C(t) = A_c \text{Cos}\,(\omega_c t) \quad\quad (8.10)$$
$$\text{Modulated Carrier :} \quad S(t) = A_c \text{Cos}\,[(\omega_c t) + \beta\ \text{Sin}\,(\omega_m t)]$$

where

- $S(t)$ = the modulated carrier
- A_c = frequency of the carrier
- ω_c = nominal frequency of the carrier frequency
- $\beta = \Delta f / f_m$ = modulation index
- $\Delta f = D_f A_m$ = frequency deviation
- D_f = constant
- A_m = amplitude of the input modulating signal

By inspecting the modulated carrier frequency, we observe that S(t) depends on both frequency and amplitude of the input signal. The spectral response is given in Fig. 8.6. Notice that the carrier frequency after modulation varies back and forth from the nominal frequency f_c as depicted in the figure. In Fig. 8.6, we see that:

- As time passes, the carrier moves back and forth in frequency in exact step with the input signal
- Frequency deviation is proportional to the input signal voltage
- A group of many side bands is created, spaced from carrier by amounts $N \times f_i$
- Relative strength of each side band depends on the Bessel function
- Strength of individual side bands far away from the carrier is proportional to (freq. deviation x input frequency)
- Higher-order spectral components are negligible
- Carson's rule can be used to determine the approximate bandwidth: bandwidth required = $2 \times$ (highest input frequency + frequency deviation)

8.3.4 Carson's Rule and FM Bandwidth

Carson's rule, a rule of thumb, states that more than 98% of the power of FM signal lies within a bandwidth given by the following approximation:

$$\text{FM Bandwidth (BW)} = 2f_m(1 + \beta) \quad\quad (8.11)$$

where

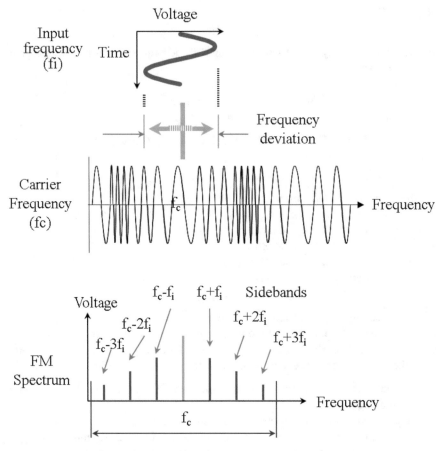

Fig. 8.6 FM spectrum. As time passes, the carrier moves back and forth in frequency in exact step with the input signal and generates an infinite number of side bands

- $\beta = \Delta f/f_m =$ modulation index
- $\Delta f =$ peak deviation of the instantaneous frequency from the center of the carrier frequency
- $f_m =$ highest frequency of the modulating signal

8.3.5 Bessel Function and FM Bandwidth

FM bandwidth can be estimated by means of the Bessel function of the first kind. For a single-tone modulation, it can be obtained as a function of the side band number and the modulation index. This is given by

For a given β, the solution for $S(t)$ is given by

$$S(t) = A_c \cos\left[\omega_c t + \beta \sin(\omega_m t)\right]$$
$$= J_0(\beta) \cos(\omega_c t)$$
$$+ \; J_1(\beta) \cos(\omega_c t + \omega_m t) + J_2(\beta) \cos(\omega_c t + 2\omega_m t) + J_3(\beta) \cos(\omega_c t + 3\omega_m t) + \ldots$$
$$- \; J_1(\beta) \cos(\omega_c t - \omega_m t) - J_2(\beta) \cos(\omega_c t - 2\omega_m t) - J_3(\beta) \cos(\omega_c t - 3\omega_m t) + \ldots$$

$$(8.12)$$

Here, J_s are the Bessel functions, representing the amplitude of the side bands. J_0 (β) is the amplitude of the fundamental spectral component, and the remaining spectral components are the side bands. Each side band is separated by the input modulating frequency. These values are also available as a standard Bessel function table.

As an example, Table 8.1 provides a few Bessel parameters to illustrate the concept. In this table, the carrier and side band amplitude powers: J_0, J_1, J_2, etc. are presented for different values of β. Here, J_0 is carrier power before modulation. After modulation, with a given value of β, power is taken from the carrier and distributed among the side bands. Also, there is a unique value of β, for which the carrier amplitude becomes zero and all the signal power is in the side bands.

Note that, in FM, the side bands are on both sides of the carrier. Therefore, the total bandwidth includes spectral components from both sides of the carrier. For a given low β, this is shown in Fig. 8.7. Here, J_0 is the spectral component of the carrier. J_1, J_2, J_3, etc. are the side bands. Each side band is separated by the input modulating frequency fm. After modulation, power is taken from the carrier J_0 and distributed among the side bands, depending on the modulation index β.

Table 8.1 Bessel function table	β	J_0	J_1	J_2	J_3	J_4	J_5
	0	1					
	0.25	0.98	0.12				
	0.5	0.94	0.24	0.03			
	1	0.77	0.44	0.11	0.02		
	1.5	0.51	0.56	0.23	0.06	0.01	
	2	0.22	0.58	0.35	0.13	0.03	

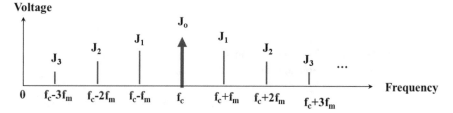

Fig. 8.7 FM side bands. Power is taken from the carrier J_0 and distributed among the side bands J_1, J_2, J_3, etc. Each side band is separated by the modulating frequency f_m

Voltage

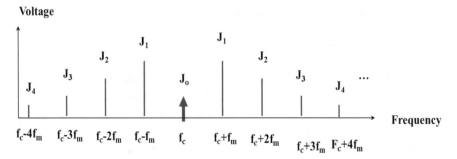

Fig. 8.8 FM side bands for large β. More power is taken from the carrier J_0 and distributed among the side bands J_1, J_2, J_3, J_4, etc. Each side band is separated by the modulating frequency f_m

Figure 8.8 shows another scenario where β is large. In this case, more power is taken from the carrier and distributed among the side band while creating more significant side bands, requiring more bandwidth.

8.3.6   FM Bandwidth Dilemma

In FM, we notice that if the modulation index is low, the occupied bandwidth is low and the side bands take less power from the carrier, making the modulation less efficient. On the other hand, if the modulation index is high, the occupied bandwidth is also high, while the side bands retain most of the power, making the modulation more efficient at the expense of bandwidth. This is a dilemma in FM. However, FM is more popular because it is less sensitive to noise.

Problem 8.1
Given:

$f_m = 1$ kHz
$f_c = 1$ MHz
$\beta = 1.$

Find:

(a) The FM bandwidth using Carson's rule
(b) The FM bandwidth using the Bessel function

Solution:

(a) FM BW (Carson's rule) $= 2f_m (1 + \beta) = 2 \times 1$ kHz $(1 + 1) = 4$ kHz
(b) For $\beta = 1$, see Table 8.1:

- $J_0 = 0.7\,7$, $J_1 = 0.44$, $J_2 = 0.11$, $J_3 = 0.02$

where J_0 is the carrier and J_1, J_2, and J_3 are the side bands. Here we can neglect J_3 since it has a negligible power. Therefore, there are two significant upper and lower side bands, and each side band is 1 kHz apart. Therefore, FM bandwidth is $2 \times$ number of significant side bands $= 2 \times 2 = 4$ kHz. See figure below:

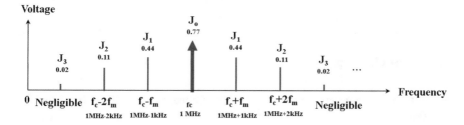

Problem 8.2
Given:

$f_m = 1$ kHz
$f_c = 1$ MHz
$\beta = 2$

Find:

(a) The FM bandwidth using Carson's rule
(b) The FM bandwidth using the Bessel function

Solution:

(a) FM BW (Carson's rule) $= 2f_m (1 + \beta) = 2 \times 1$ kHz $(1 + 2) = 6$ kHz
(b) For $\beta = 2$, see Table 8.1:

- $J_0 = 0.22$, $J_1 = 0.58$, $J_2 = 0.35$, $J_3 = 0.13$, $J_4 = 0.03$

Here, we have four upper and four lower side bands, and each side band is separated by 1 kHz. Since the J_4 is negligible, we take three upper side bands and three lower side bands. Therefore, the FM bandwidth is $2 \times 3 = 6$ kHz. See figure below:

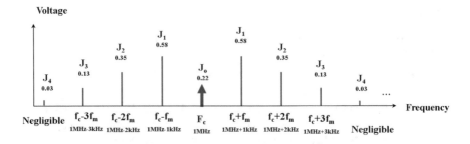

8.4   Conclusions

- This chapter presents a brief overview of frequency modulation (FM) and its attributes.
- The key concept and the underlying principle of FM are presented with numerous illustrations to make them easier for readers to grasp.
- FM spectrum is addressed with illustrations.
- FM bandwidth is estimated by means of Carson's rule and the Bessel function.
- FM bandwidth dilemma is explained and problems are given to illustrate the concept.

References

1. E.H. Armstrong, A method of reducing disturbances in radio signaling by a system of frequency modulation. Proc. IRE **24**(5), 689–740 (1936). https://doi.org/10.1109/JRPROC.1936.227383
2. B.P. Lathi, *Communication Systems* (Wiley, New York, 1968), pp. 214–217. ISBN 0-471-51832-8
3. W. Leon, I.I. Couch, *Digital and Analog Communication Systems*, 7th edn. (Prentice-Hall, Inc., Englewood Cliffs, 2001). ISBN: 0-13-142492-0
4. S. Haykin, *Communication Systems*, 4th edn. (Wiley, New York, 2001)
5. M.O. Felix, FM Systems Of Exceptional Bandwidth. Proc. IEEE **112**(9), 1664 (1965)

Chapter 9
Amplitude Shift Keying (ASK)

Topics

- Introduction
- Amplitude Shift Keying (ASK)
- ASK Spectrum
- ASK Bandwidth
- Performance Analysis

9.1 Introduction

In amplitude shift keying (ASK), the amplitude of the carrier changes in discrete levels in accordance with the input digital signal, while the frequency of the carrier remains the same. This is shown in Fig. 9.1 where:

- $m(t)$ is the input modulating digital signal
- $C(t)$ is the carrier frequency
- $S(t)$ is the ASK modulated carrier frequency

As shown in the figure, the digital binary signal changes the amplitude of the carrier in two discrete levels. This enables the receiver to extract the digital signal by demodulation. Notice that the amplitude of the carrier changes in accordance with the input signal, while the frequency of the carrier does not change after modulation. However, it can be shown that the modulated carrier $S(t)$ contains several spectral components, requiring frequency domain analysis.

In the following sections, the above disciplines in ASK modulation will be presented along with the respective spectrum and bandwidth. These materials have been augmented by diagrams and associated waveforms to make them easier for readers to grasp.

S. Faruque, *Free Space Laser Communication with Ambient Light Compensation*, https://doi.org/10.1007/978-3-030-57484-0_9

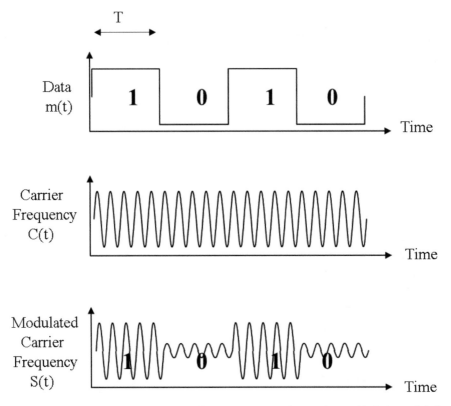

Fig. 9.1 ASK waveforms. The amplitude of the carrier changes in accordance with the input digital signal. The frequency of the carrier remains the same

9.2 ASK Modulation

Amplitude shift keying (ASK), also known as on-off keying (OOK), is a method of digital modulation that utilizes amplitude shifting of the relative amplitude of the career frequency [1–3]. The signal to be modulated and transmitted is binary, which is encoded before modulation. This is an indispensable task in digital communications, where redundant bits are added with the raw data that enables the receiver to detect and correct bit errors, if they occur during transmission [4–16].

While there are many error coding scheme available, we will use a simple coding technique, known as "block coding" to illustrate the concept.

Figure 9.2 shows an encoded ASK modulation scheme using (15, 8) block code where an 8-bit data block is formed as M-rows and N-columns ($M = 4$, $N = 2$). The product $MN = k = 8$ is the dimension of the information bits before coding. Next, a horizontal parity P_H is appended to each row, and a vertical parity P_V is appended to each column. The resulting augmented dimension is given by the product $(M + 1)$

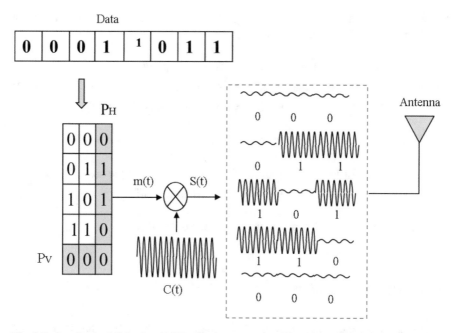

Fig. 9.2 Amplitude shift keying (ASK), also known as on-off keying (OOK). The input encoded data block is transmitted row by row. The amplitude of the carrier frequency changes in accordance with the input digital signal

$(N + 1) = n = 15$, which is then ASK modulated and transmitted row by row. The rate of this coding scheme is given by

$$\text{Code Rate} : r = (MN)/[(M + 1)(N + 1)] = (4 \times 2)/(5 \times 3) = 8/15 \qquad (9.1)$$

The coded bit rate R_{b2} is given by

$$R_{b2} = \text{Uncoded Bit Rate/Code Rate} = R_{b1}/r = R_{b1}(15/8) \qquad (9.2)$$

Next, the coded bits are modulated by means of the ASK modulator as shown in the figure. Here:

- The input digital signal is the encoded bit sequence we want to transmit.
- Carrier is the radio frequency without modulation.
- Output is the ASK modulated carrier, which has two amplitudes corresponding to the binary input signal. For binary signal 1, the carrier is *on*. For the binary signal 0, the carrier is *off*; however, a small residual signal may remain due to noise, interference, etc. as indicated in the figure.

As shown in the figure, the amplitude of the carrier changes in discrete levels, in accordance with the input signal, where:

$$\text{Input Data} : m(t) = 0 \text{ or } 1 \text{ (coded data)}$$
$$\text{Carrier Frequency} : C(t) = A \cos(\omega t) \tag{9.3}$$
$$\text{Modulated Carrier} : S(t) = m(t)C(t) = m(t)A\cos(\omega t)$$

Therefore,

$$\text{For } m(t) = 1 : S(t) = A\cos(\omega t) \text{ i.e.the carrier is on}$$
$$\text{For } m(t) = 0 : S(t) = 0, \text{i.e.the carrier is off} \tag{9.4}$$

where A is the amplitude and ω is the frequency of the carrier.

9.3   ASK Demodulation

Once the modulated binary data has been transmitted, it needs to be received and demodulated. This is often accomplished with the use of a bandpass filter. In the case of ASK, the receiver needs to utilize one bandpass filter that is tuned to the appropriate carrier frequency. As the signal enters the receiver, it passes through the filter, and a decision as to the value of each bit is made to recover the encoded data block, along with horizontal and vertical parities. Next, the receiver appends horizontal and vertical parities $P_H{}^*$ and $P_V{}^*$ to check parity failures and recovers the data block. This is shown in Fig. 3.7 having no errors. If there is an error, there will be a parity failure in $P_H{}^*$ and $P_V{}^*$ to pinpoint the error (Fig. 9.3).

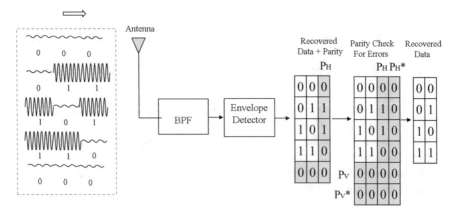

Fig. 9.3 Data recovery process in ASK, showing no errors. If there is an error, there will be a parity failure in $P_H{}^*$ and $P_V{}^*$ to pinpoint the error

9.4   ASK Bandwidth

In wireless communications, the scarcity of RF spectrum is well-known. For this reason, we have to be vigilant about using transmission bandwidth in error control coding and modulation. The transmission bandwidth depends on:

- Spectral response of the encoded data
- Spectral response of the carrier frequency
- Modulation type

9.4.1   Spectral Response of the Encoded Data

In digital communications, data is generally referred to as a nonperiodic digital signal. It has two values:

- Binary-1 = high, period = T
- Binary-0 = low, period = T

Also, data can be represented in two ways:

- Time domain representation
- Frequency domain representation

The time domain representation (Figure 9.4a), known as non-return-to-zero (NRZ), is given by

$$V(t) = V \qquad < 0 < t < T$$
$$= 0 \qquad \text{elsewhere} \tag{9.5}$$

The frequency domain representation is given by "Fourier transform":

$$V(\omega) = \int_0^T V.e^{-j\omega t} dt \tag{9.6}$$

$$|V(\omega)| = VT \left[\frac{\text{Sin}(\omega T/2)}{\omega T/2} \right]$$

$$P(\omega) = \left(\frac{1}{T} \right) |V(\omega)|^2 = V^2 T \left[\frac{\text{Sin}(\omega T/2)}{\omega T/2} \right]^2 \tag{9.7}$$

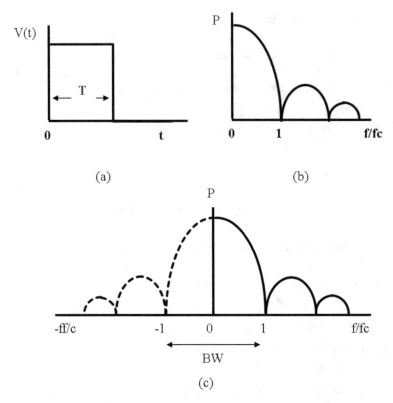

Fig. 9.4 (**a**) Discrete time digital signal. (**b**) Its one-sided power spectral density and (**c**) two-sided power spectral density. The bandwidth associated with the non-return-to-zero (NRZ) data is 2Rb, where R_b is the bit rate

Here, $P(\omega)$ is the power spectral density. This is plotted in Figure 9.4b. The main lobe corresponds to the fundamental frequency and side lobes correspond to harmonic components. The bandwidth of the power spectrum is proportional to the frequency. In practice, the side lobes are filtered out since they are relatively insignificant with respect to the main lobe. Therefore, the one-sided bandwidth is given by the ratio $f/f_b = 1$. In other words, the one-sided bandwidth $= f = f_b$, where $f_b = R_b = 1/T$, T being the bit duration.

The general equation for two-sided response is given by

$$V(\omega) = \int_{-\infty}^{\infty} V(t).e^{-j\omega t} dt$$

In this case, $V(\omega)$ is called two-sided spectrum of $V(t)$. This is due to both positive and negative frequencies used in the integral. The function can be a voltage or a current. Figure 9.4c shows the two-sided response, where the bandwidth is determined by the main lobe as shown below:

$$\text{Two sided bandwidth (BW)} = 2R_b(R_b = \text{Bit rate before coding}) \qquad (9.8)$$

Important Notes:
1. If R_b is the bit rate before coding, and if the data is NRZ, then the bandwidth associated with the raw data will be $2R_b$. For example, if the bit rate before coding is 10 kb/s, then the bandwidth associated with the raw data will be $2 \times 10\text{kb/s} = 20$ kHz.
2. If R_b is the bit rate before coding, the code rate is r, and if the data is NRZ, then the bitrate after coding will be $R_b(\text{coded}) = R_b(\text{uncoded})r$. The corresponding bandwidth associated with the coded data will be $2R_b$ (coded) $= 2R_b$ (uncoded)/r. For example, if the bit rate before coding is 10 kb/s and the code rate $r = 1/2$, the coded bit rate will be R_b (coded) $= R_b$ (uncoded)/$r = 10/0.5 = 20$ kb/s. The corresponding bandwidth associated with the coded data will be $2 \times 20 = 40$ kHz.

9.4.2 Spectral Response of the Carrier Frequency Before Modulation

A carrier frequency is essentially a sinusoidal waveform, which is periodic and continuous with respect to time. It has one frequency component. For example, the sine wave is described by the following time domain equation:

$$V(t) = V_p \text{Sin}(\omega t_c) \qquad (9.9)$$

where

Vp = peak voltage

- $\omega_c = 2\pi f_c$
- f_c = carrier frequency in Hz

Figure 9.5 shows the characteristics of a sine wave and its spectral response. Since the frequency is constant, its spectral response is located in the horizontal axis, and the peak voltage is shown in the vertical axis. The corresponding bandwidth is zero.

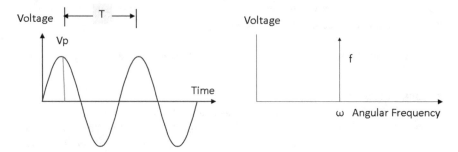

Fig. 9.5 High-frequency carrier frequency response

9.4.3 ASK Bandwidth at a Glance

In ASK, the amplitude of the carriers changes in discrete levels, in accordance with the input signal, where:

- Input data: $m(t) = 0$ or 1
- Carrier frequency: $C(t) = A_c \, \mathrm{Cos}(\omega_c t)$
- Modulated carrier: $S(t) = m(t)C(t) = m(t)A_c \, \mathrm{Cos}(\omega_c t)$

Since $m(t)$ is the input digital signal and it contains an infinite number of harmonically related sinusoidal waveforms and that we keep the fundamental and filter out the higher-order components, we write

$$m(t) = A_m \mathrm{Sin}\,(\omega_m t)$$

The ASK modulated signal then becomes

$$\begin{aligned} S(t) &= m(t)S(t) = A_m A_c \mathrm{Sin}(\omega_m t)\mathrm{Cos}(\omega_c t) \\ &= 1/2\, A_m A_c \left[\mathrm{Sin}(\omega_c - \omega_m)t + \mathrm{Sin}(\omega_c \pm \omega_m)t \right] \end{aligned} \tag{9.10}$$

The spectral response is depicted in Fig. 3.14. Notice that the spectral response after ASK modulation is the shifted version of the NRZ data. Bandwidth is given by (Fig. 9.6)

$\mathrm{BW} = 2R_b$ (coded), where R_b is the coded bit rate

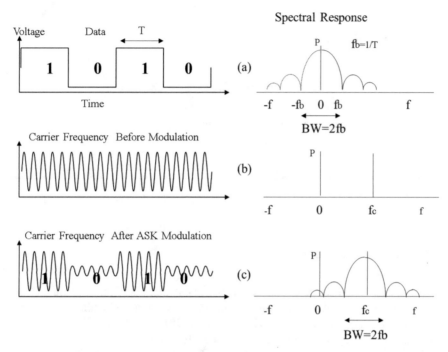

Fig. 9.6 ASK bandwidth at a glance. (**a**) Spectral response of NRZ data before modulation. (**b**) Spectral response of the carrier before modulation. (**c**) Spectral response of the carrier after modulation. The transmission bandwidth is $2f_b$, where f_b is the bit rate and $T = 1/f_b$ is the bit duration for NRZ data

---

Problem 1

Given:

- Bit rate before coding: $R_{B1} = 10$ kb/s
- Code rate: $r = 8/15$
- Modulation: ASK

Find:

(a) The bit rate after coding: R_{B2}
(b) Transmission bandwidth: BW

Solution:

(a) Bit rate after coding (R_{B2}):

$$R_{B2} = R_{B1}/r = 10 \text{ kb/s } (15/8) = 18.75 \text{ kb/s}$$

(b) Transmission BW $= 2 \times 18.75$ kb/s $= 37.5$ kHz

9.5   BER Performance

It is well-known that an (n, k) block code, where k = number of information bits and n = number of coded bits, can correct t errors [4, 5]. A measure of coding gain is then obtained by comparing the uncoded word error, WER_U, to the coded word WER_C. We examine this by means of the following analytical means:

Let the uncoded word error be defined as WER_U. Then, with ASK modulation, the uncoded BER will be given by

$$BER_U = 0.5 \text{ EXP}(-E_B/2N_o) \tag{9.11}$$

The probability that the uncoded word (WER_U) will be received in error is 1 minus the product of the probabilities that each bit will be received correctly. Thus, we write

$$WER_U = 1 - (1 - BER_U)^k \tag{9.12}$$

Let the coded word error be defined as WER_C. Since $n > k$, the coded bit energy to noise ratio will be modified to Ec/No, where Ec/No = Eb/No +10log(k/n). Therefore, the coded BERc will be

$$BERc = 0.5 \text{ EXP } (-E_c/2N_o) \tag{9.13}$$

The corresponding coded word error rate is

$$WER_C = \sum_{k=t+1}^{n} \binom{n}{k} BER_C^K (1 - BER_C)^{n-k} \tag{9.14}$$

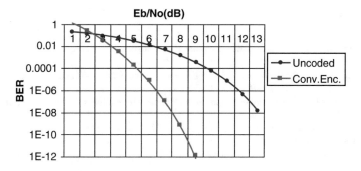

Fig. 9.7 Typical error performance in AWGN

When BERc <0.5, the first term in the summation is the dominant one; therefore, Eq. (XXX) can be simplified as

$$\text{WER}_C \approx \binom{n}{k} \text{BER}_c^k (1 - BER_c)^{n-k} \tag{9.15}$$

Using (15, 8) block code ($n = 15$, $k = 8$, $t = 1$), we obtain the coded and the uncoded WER as shown in Fig. 9.7. Coding gain is the difference in Eb/No between the two curves. Notice that at least 3 to 4 db coding gain is available in this example where $r = k/n = 8/15$.

9.6   Conclusions

- This chapter presents ASK modulation and its attributes.
- Numerous illustrations are provided to show how amplitude of the carrier changes in discrete levels in accordance with the input digital signal, while the frequency of the carrier remains the same.
- The Fourier transform is used to derive the spectral components and ASK bandwidth is calculated.
- Bit error rate (BER) performance is presented.
- The materials have been augmented by diagrams and associated waveforms to make them easier for readers to grasp.

References

1. D.R. Smith, *Digital Transmission Systems* (Van Nostrand Reinhold Co, New York, 1985) ISBN: 0442009178,
2. W. Leon, I.I. Couch, *Digital and Analog Communication Systems*, 7th edn. (Prentice-Hall, Inc, Englewood Cliffs, 2001) ISBN: 0-13-142492-0
3. B. Sklar, *Digital Communications Fundamentals and Applications* (Prentice Hall, Englewood Cliffs, 1988)
4. S. Faruque, *Radio Frequency Channel Coding Made Easy* (Springer, Cham, 2016) ISBN: 978-3-319-21169-5
5. C.C. George et al., *Error Correction Coding for Digital Communications* (Plenum Press, New York, 1981)
6. G. Ungerboeck, Channel coding with multilevel/multiphase signals. IEEE Trans. Inf. Theory **IT28**, 55–67 (January 1982)
7. S. Lin, D.J. Costello Jr., *Error Control Coding: Fundamentals and Applications* (Prentice-Hall, Inc., Englewood Cliffs, 1983)
8. R.E. Blahut, *Theory and Practice of Error Control Codes* (Addison-Wesley, Reading, 1983)
9. J.H. van Lint, *Introduction to Coding Theory*, GTM 86, 2nd edn. (Springer-Verlag, New York, 1992), p. 31. ISBN 3-540-54894-7
10. F.J. Mac Williams, N.J.A. Sloane, *The Theory of Error-Correcting Codes* (North-Holland, Amsterdam, 1977), p. 35. ISBN 0-444-85193-3
11. W. Huffman, V. Pless, *Fundamentals of Error-Correcting Codes* (Cambridge University Press, Cambridge, 2003) ISBN 978-0-521-78280-7
12. P.M. Ebert, S.Y. Tong, Convolutional Reed-Soloman codes. Bell Syst. Tech. **48**, 729–742 (1968)
13. R.G. Gallager, *Information Theory and Reliable Communications* (Wiley, New York, 1968)
14. A. Kohlenbero, G.D. Forney Jr., Convolutional coding for channels with memory. IEEE Trans. Inf. Theory **IT-14**, 618–626 (1968)
15. M.A. Reddy, S. M, Further results on convolutional codes derived from block codes. Inf. Control. **13**, 357–362 (1968)
16. S.M. Reddy, *A Class of Linear Convolutional Codes for Compound Channels*, Technical Report (Bell Telephone Laboratories, Holmdel, 1968)

Chapter 10
Frequency Shift Keying (FSK)

Topics
- Introduction
- Frequency Shift Keying (FSK)
- FSK Spectrum
- FSK Bandwidth
- Performance Analysis

10.1   Introduction

In frequency shift keying (FSK), the frequency of the carrier changes in discrete levels in accordance with the input digital signal, while the amplitude of the carrier remains the same. This is shown in Fig. 10.1 where:

- $m(t)$ is the input modulating digital signal
- $C(t)$ is the carrier frequency
- $S(t)$ is the FSK modulated carrier frequency

As shown in the figure, the digital binary signal changes the frequency of the carrier on two discrete levels. This enables the receiver to extract the digital signal by demodulation. Notice that the frequency of the carrier changes in accordance with the input signal, while the amplitude of the carrier does not change after modulation. However, it can be shown that the modulated carrier S(t) contains several spectral components, requiring frequency domain analysis.

In the following sections, the above disciplines in FSK modulation will be presented, along with the respective spectrum and bandwidth. These materials have been augmented by diagrams and associated waveforms to make them easier for readers to grasp.

© The Editor(s) (if applicable) and The Author(s), under exclusive license to
Springer Nature Switzerland AG 2021
S. Faruque, *Free Space Laser Communication with Ambient Light Compensation*,
https://doi.org/10.1007/978-3-030-57484-0_10

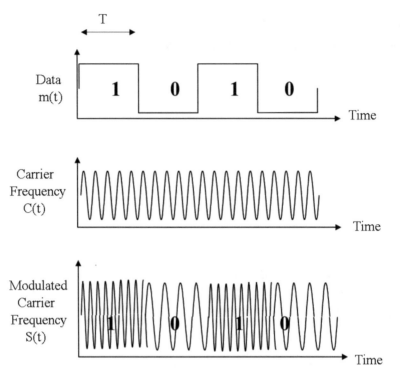

Fig. 10.1 FSK waveforms. The frequency of the carrier changes in accordance with the input digital signal. The amplitude of the carrier remains the same

10.2 Frequency Shift Keying (FSK) Modulation

Frequency shift keying (FSK) is a method of digital modulation that utilizes frequency shifting of the relative frequency content of the signal [1–3]. The signal to be modulated and transmitted is binary, which is encoded before modulation. This is an indispensable task in digital communications, where redundant bits are added with the raw data that enables the receiver to detect and correct bit errors, if they occur during transmission [4–16]. While there are many error coding schemes available, we will use a simple coding technique, known as "block coding," to illustrate the concept.

Figure 10.2 shows an encoded FSK modulation scheme using (15, 8) block code where an 8-bit data block is formed as M-rows and N-columns ($M = 4, N = 2$). The product $MN = k = 8$ is the dimension of the information bits before coding. Next, a horizontal parity P_H is appended to each row, and a vertical parity P_V is appended to each column. The resulting augmented dimension is given by the product $(M + 1)$ $(N + 1) = n = 15$, which is then FSK modulated and transmitted row by row. The rate of this coding scheme is given by

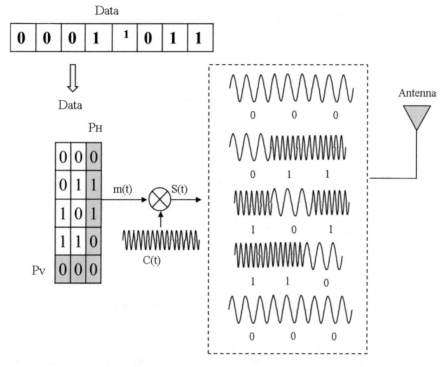

Fig. 10.2 Binary frequency shift keying (BFSK) modulation. The input encoded data block is transmitted row by row. The frequency of the carrier changes in accordance with the input digital signal

$$\text{Code Rate} : r = (MN)/[(M + 1)(N + 1)] = (4 \times 2)/(5 \times 3) = 8/15 \quad (10.1)$$

The coded bit rate R_{b2} is given by

$$R_{b2} = \text{Uncoded Bit Rate/Code Rate} = R_{b1}/r = R_{b1}(15/8) \quad (10.2)$$

Next, the coded bits are modulated by means of the FSK modulator as shown in the figure. Here:

- The input digital signal is the encoded bit sequence we want to transmit
- The carrier is the radio frequency without modulation
- The output is the FSK modulated carrier, which has two frequencies corresponding to the binary input signal
- For binary signal 1, the carrier changes to $f_c - \Delta f$
- For binary signal 0, the carrier changes to $f_c + \Delta f$
- The total frequency deviation is $2\Delta f$

As shown in Fig. 3.8, the frequency of the carrier changes in discrete levels, in accordance with the input signals. We have:

$$\text{Input Data} : m(t) = 0 \text{ or } 1$$
$$\text{Carrier Frequency} : C(t) = A \ \text{Cos} \ (\omega t)$$
$$\text{Modulated Carrier} : S(t) = A \text{Cos}(t)t, \text{For } m(t) = 1 \tag{10.3}$$
$$S(t) = A \text{Cos} \ (\omega + \Delta\omega)t, \text{For } m(t) = 0$$

where

$$A = \text{Frequency of the carrier}$$
$$\omega = \text{Nominal frequency of the carrier frequency} \tag{10.4}$$
$$\Delta\omega = \text{Frequency deviation}$$

10.3   Frequency Shift Keying (FSK) Demodulation

Once the modulated binary data has been transmitted, it needs to be received and demodulated. This is often accomplished by the use of bandpass filters. In the case of binary FSK, the receiver needs to utilize two bandpass filters that are tuned to the appropriate frequencies. Since the nominal carrier frequency and the frequency deviation are known, this is relatively straightforward. One bandpass filter will be centered at the frequency ω_1 and the other at ω_2. As the signal enters the receiver, it passes through the filters and a decision as to the value of each bit is made. This is shown in Fig. 10.3. In order to assure that the bits are decoded correctly, the frequency deviation needs to be chosen with the limitations of the filters in mind to eliminate crossover.

10.4   FSK Bandwidth

In wireless communications, the scarcity of RF spectrum is well-known. For this reason, we have to be vigilant about using transmission bandwidth in error control coding and modulation. The transmission bandwidth depends on:

- Spectral response of the encoded data
- Spectral response of the carrier frequency
- Modulation type

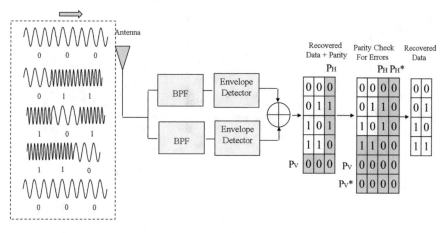

Fig. 10.3 Binary FSK detector utilizing two matched bandpass filters

10.4.1 Spectral Response of the Encoded Data

In digital communications, data is generally referred to as a nonperiodic digital signal. It has two values:

- Binary-1 = high, period = T
- Binary-0 = low, period = T

Also, data can be represented in two ways:

- Time domain representation
- Frequency domain representation

The time domain representation (Figure 10.4a), known as non-return-to-zero (NRZ), is given by

$$V(t) = V \quad < 0 < t < T$$
$$= 0 \quad \text{elsewhere} \tag{10.5}$$

The frequency domain representation is given by "Fourier transform":

$$V(\omega) = \int_0^T V.e^{-j\omega t} dt \tag{10.6}$$

$$|V(\omega)| = VT\left[\frac{\text{Sin}(\omega T/2)}{\omega T/2}\right]$$

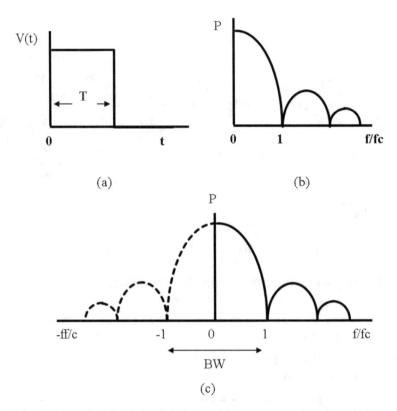

Fig. 10.4 (a) Discrete time digital signal. (b) Its one-sided power spectral density and (c) two-sided power spectral density. The bandwidth associated with the non-return-to-zero (NRZ) data is 2Rb, where R_b is the bit rate

$$P(\omega) = \left(\frac{1}{T}\right)|V(\omega)|^2 = V^2 T \left[\frac{\mathrm{Sin}(\omega T/2)}{\omega T/2}\right]^2 \tag{10.7}$$

Here, $P(\omega)$ is the power spectral density. This is plotted in Figure 10.4b. The main lobe corresponds to the fundamental frequency, while side lobes correspond to harmonic components. The bandwidth of the power spectrum is proportional to the frequency. In practice, the side lobes are filtered out, since they are relatively insignificant with respect to the main lobe. Therefore, the one-sided bandwidth is given by the ratio $f/fb = 1$. In other words, the one-sided bandwidth $= f = f_b$, where $f_b = R_b = 1/T$, T being the bit duration.

The general equation for two-sided response is given by

$$V(\omega) = \int_{-\infty}^{\infty} V(t).e^{-j\omega t} dt$$

In this case, $V(\omega)$ is called the two-sided spectrum of $V(t)$. This is due to both positive and negative frequencies used in the integral. The function can be either a voltage or a current. Figure 10.4c shows the two-sided response, where the bandwidth is determined by the main lobe as shown below:

$$\text{Two sided bandwidth (BW)} = 2R_b (R_b = \text{Bit rate before coding}) \quad (10.8)$$

Important Notes
1. If R_b is the bit rate before coding, and if the data is NRZ, then the bandwidth associated with the raw data will be $2R_b$. For example, if the bit rate before coding is 10 kb/s, then the bandwidth associated with the raw data will be 2×10kb/s $= 20$ kHz.
2. If R_b is the bit rate before coding, the code rate is r, and the data is NRZ, then the bitrate after coding will be Rb(coded) $= R_b$(uncoded)/r. The corresponding bandwidth associated with the coded data will be $2R_b$ (coded) $= 2Rb$ (uncoded)/r. For example, if the bit rate before coding is 10 kb/s and the code rate r $= 1/2$, the coded bit rate will be R_b (coded) $= R_b$ (uncoded)/r $= 10/0.5 = 20$ kb/s. The corresponding bandwidth associated with the coded data will be $2 \times 20 = 40$ kHz.

10.4.2   Spectral Response of the Carrier Frequency Before Modulation

A carrier frequency is essentially a sinusoidal waveform, which is periodic and continuous with respect to time. It has one frequency component. For example, the sine wave is described by the following time domain equation:

$$V(t) = V_p \text{Sin}(\omega t_c) \quad (10.9)$$

where

Vp $=$ peak voltage

- $\omega_c = 2\pi f_c$
- $f_c =$ carrier Frequency in Hz

Figure 10.5 shows the characteristics of a sine wave and its spectral response. Since the frequency is constant, its spectral response is located in the horizontal axis, and the peak voltage is shown in the vertical axis. The corresponding bandwidth is zero.

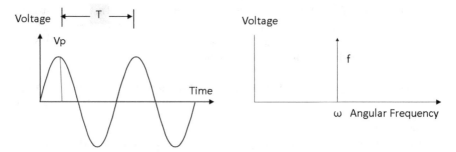

Fig. 10.5 High-frequency carrier response

10.4.3   FSK Bandwidth at a Glance

In FSK, the frequency of the carrier changes in two discrete levels, in accordance with the input signals. We have:

- Input data: m(t) = 0 or 1
- Carrier frequency: $C(t) = A \operatorname{Cos}(\omega t)$
- Modulated carrier : $S(t) = A \operatorname{Cos}(\omega - \Delta\omega)t, \quad \text{for} \quad m(t) = 1$

$$S(t) = A \operatorname{Cos}(\omega + \Delta\omega)t, \text{for} m(t) = 0$$

where

- $S(t)$ = the modulated carrier
- A = amplitude of the carrier
- ω = nominal frequency of the carrier
- $\Delta\omega$ = frequency deviation

   The spectral response is depicted in Fig. 10.6. Notice that the carrier frequency after FSK modulation varies back and forth from the nominal frequency f_c by $\pm \Delta f_c$, where Δfc is the frequency deviation. The FSK bandwidth is given by

$$\begin{aligned} \text{BW} &= 2(f_b + \Delta f_c) \\ &= 2 f_b(1 + \Delta f_c/f_b) \\ &= 2 f_b(1 + \beta) \end{aligned} \qquad (10.10)$$

where $\beta = \Delta f/f_b$ is known as the modulation index and f_b is the coded bit frequency (bit rate Rb). The above equation is also known as "Carson's rule."

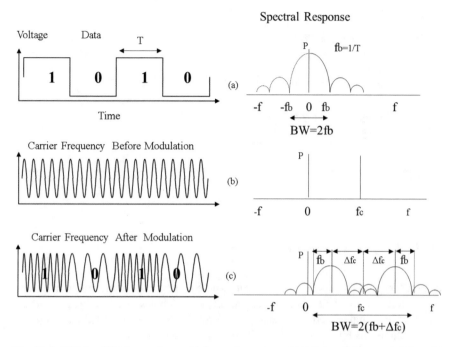

Fig. 10.6 FSK bandwidth at a glance. (**a**) Spectral response of NRZ data before modulation. (**b**) Spectral response of the carrier before modulation. (**c**) Spectral response of the carrier after modulation. The transmission bandwidth is $2(f_b + \Delta f_c)$. f_b is the bit rate and Δf_c is the frequency deviation $= 1/f_b$ which is the bit duration for NRZ data

Problem:

Given:

- Bit rate before coding: $R_{B1} = 10$ kb/s
- Code rate: $r = 8/15$
- Modulation: FSK
- Modulation index $\beta = 1$

Find:

(a) The bit rate after coding: R_{B2}
(b) Transmission bandwidth: BW

Solution:

(a) Bit rate after coding (R_{B2}):

$$R_{B2} = R_{B1}/r = 10\text{kb/s} \ (15/8) = 18.75 \ \text{kb/s}$$

(b)
Transmission BW $= 2 \ R_{B2}(1 + \beta)$

$= 2 \times 18.75\text{kb/s}(1 + 1) = 75\text{kHz}$

10.5   BER Performance

It is well-known that an (n, k) block code, where k = number of information bits and n = number of coded bits, can correct t errors [4, 5]. A measure of coding gain is then obtained by comparing the uncoded word error, WER_U, to the coded word WER_C. We examine this by means of the following analytical means:

Let the uncoded word error be defined as WER_U. Then, with FSK modulation, the uncoded BER will be given by

$$BER_U = 0.5 \ EXP(-E_B/2N_o) \tag{10.11}$$

The probability that the uncoded word (WER_U) will be received in error is 1 minus the product of the probabilities that each bit will be received correctly. Thus, we write

$$WER_U = 1 - (1 - BER_U)^k \tag{10.12}$$

Let the coded word error be defined as WER_C. Since $n > k$, the coded bit energy to noise ratio will be modified to Ec/No, where Ec/No $=$ Eb/No + 10 log(k/n). Therefore, the coded BERc will be

$$BERc = 0.5 \ EXP \ (-E_c/2N_o) \tag{10.13}$$

The corresponding coded word error rate is

$$WER_C = \sum_{k=t+1}^{n} \binom{n}{k} BER_C^K (1 - BER_C)^{n-k} \tag{10.14}$$

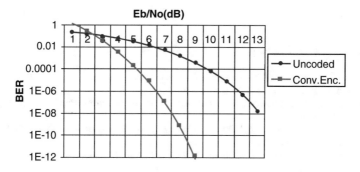

Fig. 10.7 Typical error performance in AWGN

When BERc <0.5, the first term in the summation is the dominant one; therefore, the equation can be simplified as

$$\text{WER}_C \approx \binom{n}{k} \text{BER}_c^k (1 - \text{BER}_c)^{n-k} \tag{10.15}$$

Using (15, 8) block code ($n = 15$, $k = 8$, $t = 1$), we obtain the coded and the uncoded WER as shown in Fig. 10.7. Coding gain is the difference in Eb/No between the two curves. Notice that at least 3 to 4 db coding gain is available in this example where r = k/n = 8/15.

10.6 Conclusions

- This chapter presents FSK modulation and its attributes.
- Numerous illustrations are provided to show how frequency of the carrier changes in discrete levels in accordance with the input digital signal, while the amplitude of the carrier remains the same.
- The Fourier transform is used to derive the spectral components and FSK bandwidth is calculated.
- Bit error rate (BER) performance is presented.
- These materials have been augmented by diagrams and associated waveforms to make them easier for readers to grasp.

References

1. D.R. Smith, *Digital Transmission Systems* (Van Nostrand Reinhold Co, New York, 1985) ISBN: 0442009178,
2. W. Leon, I.I. Couch, *Digital and Analog Communication Systems*, 7th edn. (Prentice-Hall, Inc, Englewood Cliffs, 2001) ISBN: 0-13-142492-0
3. B. Sklar, *Digital Communications Fundamentals and Applications* (Prentice Hall, Englewood Cliffs, 1988)
4. S. Faruque, *Radio Frequency Channel Coding Made Easy* (Springer, Cham, 2016) ISBN: 978-3-319-21169-5
5. C.C. George et al., *Error Correction Coding for Digital Communications* (Plenum Press, New York, 1981)
6. G. Ungerboeck, Channel coding with multilevel/multiphase signals. IEEE Trans. Inf. Theory **IT28**, 55–67 (January 1982)
7. S. Lin, D.J. Costello Jr., *Error Control Coding: Fundamentals and Applications* (Prentice-Hall, Inc., Englewood Cliffs, 1983)
8. R.E. Blahut, *Theory and Practice of Error Control Codes* (Addison-Wesley, Reading, 1983)
9. J.H. van Lint, *Introduction to Coding Theory*, GTM 86, 2nd edn. (Springer-Verlag, New York, 1992), p. 31. ISBN 3-540-54894-7
10. F.J. Mac Williams, N.J.A. Sloane, *The Theory of Error-Correcting Codes* (North-Holland, Amsterdam, 1977), p. 35. ISBN 0-444-85193-3
11. W. Huffman, V. Pless, *Fundamentals of Error-Correcting Codes* (Cambridge University Press, Cambridge, 2003) ISBN 978-0-521-78280-7
12. P.M. Ebert, S.Y. Tong, Convolutional Reed-Soloman codes. Bell Syst. Tech, 729–742 (1968)
13. R.G. Gallager, *Information Theory and Reliable Communications* (Wiley, New York, 1968)
14. A. Kohlenbero, G.D. Forney Jr., Convolutional coding for channels with memory. IEEE Trans. Inf. Theory **IT-14**, 618–626 (1968)
15. M.A. Reddy, S. M, Further results on convolutional codes derived from block codes. Inf. Control. **13**, 357–362 (1968)
16. S.M. Reddy, *A Class of Linear Convolutional Codes for Compound Channels*, Technical Report (Bell Telephone Laboratories, Holmdel, 1968)

Chapter 11
Phase Shift Keying (PSK)

Topics
- Introduction
- Binary Phase Shift Keying (BPSK) Modulation
- QPSK Modulation
- 8PSK Modulation
- 16PSK Modulation
- PSK Spectrum and Bandwidth

11.1 Introduction

In phase shift keying (PSK), the phase of the carrier changes in discrete levels in accordance with the input digital signal, while the amplitude of the carrier remains the same. This is shown in Fig. 11.1 where:

- $m(t)$ is the input modulating digital signal
- $C(t)$ is the carrier frequency
- $S(t)$ is the PSK modulated carrier frequency

As shown in the figure, the digital binary signal changes the phase of the carrier on two discrete levels. This enables the receiver to extract the digital signal by demodulation. Notice that the phase of the carrier changes in accordance with the input signal, while the amplitude of the carrier does not change after modulation. However, it can be shown that the modulated carrier S(t) contains several spectral components, requiring frequency domain analysis.

© The Editor(s) (if applicable) and The Author(s), under exclusive license to
Springer Nature Switzerland AG 2021
S. Faruque, *Free Space Laser Communication with Ambient Light Compensation*,
https://doi.org/10.1007/978-3-030-57484-0_11

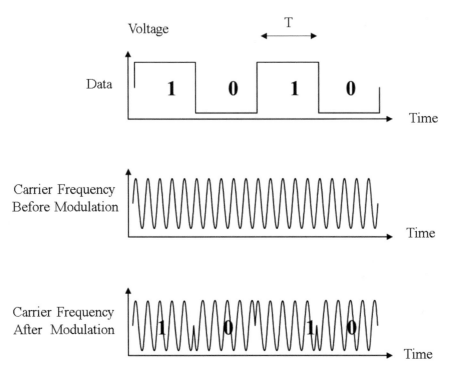

Fig. 11.1 Binary PSK (BPSK) waveforms. The phase of the carrier changes in accordance with the input digital signal. The amplitude of the carrier remains the same

In the following sections, the above disciplines in PSK modulation will be presented, along with the respective spectrum and bandwidth. These materials have been augmented by diagrams and associated waveforms to make them easier for readers to grasp.

11.2   Binary Phase Shift Keying (BPSK)

11.2.1   BPSK Modulation

Phase shift keying (PSK) is a method of digital modulation that utilizes phase shifting of the relative phase content of the signal [1–3]. The signal to be modulated and transmitted is binary, which is encoded before modulation. This is an indispensable task in digital communications, where redundant bits are added with the raw data that enables the receiver to detect and correct bit errors, if they occur during transmission [4–16]. While there are many error coding schemes available, we will use a simple coding technique, known as "block coding," to illustrate the concept.

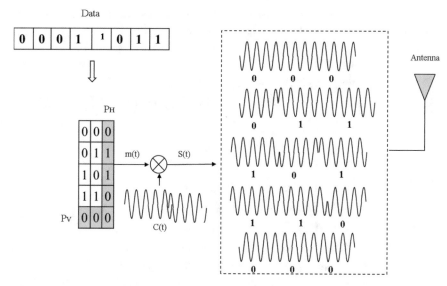

Fig. 11.2 Binary phase shift keying (BPSK) modulation. The input encoded data block is transmitted row by row. The phase of the carrier changes in accordance with the input digital signal

Figure 11.2 shows an encoded BPSK modulation scheme using (15, 8) block code where an 8-bit data block is formed as M-rows and N-columns ($M = 4, N = 2$). The product $MN = k = 8$ is the dimension of the information bits before coding. Next, a horizontal parity P_H is appended to each row, and a vertical parity P_V is appended to each column. The resulting augmented dimension is given by the product $(M + 1)(N + 1) = n = 15$, which is then PSK modulated and transmitted row by row. The rate of this coding scheme is given by

$$\text{Code Rate} : r = (MN)/[(M + 1)(N + 1)] = (4 \times 2)/(5 \times 3) = 8/15 \quad (11.1)$$

The coded bit rate R_{b2} is given by

$$R_{b2} = \text{Uncoded Bit Rate/Code Rate} = R_{b1}/r = R_{b1}(15/8) \quad (11.2)$$

Next, the coded bits are modulated by means of the PSK modulator as shown in the figure. Here:

- The input digital signal is the encoded bit sequence we want to transmit
- The carrier is the radio frequency without modulation
- The output is the PSK modulated carrier, which has two phases corresponding to the binary input signals
- For binary signal 0, $\varphi) = 0^0$
- For binary signal 1, $\varphi) = 180^0$

As shown in Fig. 11.2, the phase of the carrier changes in two discrete levels, in accordance with the input signals. Here, we have:

$$\text{Input Data} : m(t) = 0 \text{ or } 1$$
$$\text{Carrier Frequency} : C(t) = A \cos(\omega_{ct})$$
$$\text{Modulated Carrier} : S(t) = A_c \cos[\omega_c t + 2\pi/M) m(t)] \ m(t) = 0, 1, 2, 3, \ldots M - 1$$

$$(11.3)$$

where

$$A_c = \text{amplitude of the carrier frequency}$$
$$\omega_c = \text{angular frequency of the carrier}$$
$$M = 2, 4, 8, 16 \ldots$$
$$\text{In BPSK, there are two phases, 1 bit/phase } (M = 2) \qquad (11.4)$$
$$\text{In QPSK, there are four phases, 2-bits/phase, } M = 4$$
$$\text{In 8PSK, there are 8 phases, 3-bits/phase, } \quad M = 8$$
$$\text{In 16PSK, there are 16 phases, 4-bits/phase, } \quad M = 16$$

We can also represent the BPSK modulator as a signal constellation diagram with $M = 2$, 1 bit per phase. This is shown in Fig. 11.3, where the input raw data, having a bit rate R_{b1}, is encoded by means of a rate r encoder. The encoded data, having a bit rate $R_{b2} = R_{b1}/r \ (r < 1)$, is modulated by the BPSK modulator as shown in Fig. 11.3.

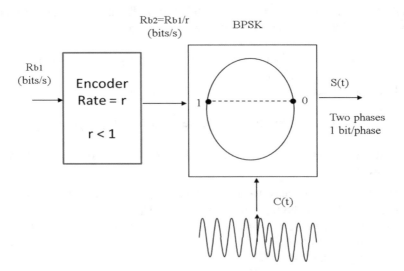

Fig. 11.3 BPSK signal constellation having two symbols, 1 bit per symbol

The BPSK modulator takes 1 bit at a time to construct the phase constellation having two phases, also known as "symbols," where each symbol represents 1 bit. The symbol rate is therefore the same as the encoded bit rate R_{b2}.

Therefore, the BPSK modulator has the following specifications:

- Two phases or two symbols
- 1 bit/symbol

The above specifications govern the transmission bandwidth, as we shall see later.

11.2.2   BPSK Demodulation

Once the modulated binary data has been transmitted, it needs to be received and demodulated. This is often accomplished with the use of a phase detector, typically known as phase-locked loop (PLL). As the signal enters the receiver, it passes through the PLL. The PLL locks the incoming carrier frequency and tracks the variations in frequency and phase. This is known as coherent detection technique, where the knowledge of the carrier frequency and phase must be known to the receiver.

Figure 11.4 shows a simplified diagram of a BPSK demodulator along with the data recovery process. In order to assure that the bits are decoded correctly, the phase deviation needs to be chosen with the limitations of the PLL in mind to eliminate crossover.

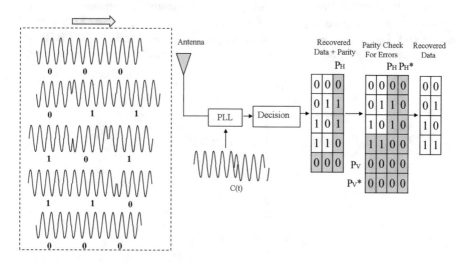

Fig. 11.4 Binary PSK detector showing data recovery process

11.3   QPSK Modulation

In QPSK, the input raw data, having a bit rate R_{b1}, is encoded by a rate r $(r < 1)$ encoder. The encoded data, having a bit rate $R_{b2} = R_{b1}/r$, is serial to parallel converted into two parallel streams. The encoded bit rate, now reduced in speed by a factor of 2, is modulated by the QPSK modulator as shown in Fig. 11.5.

The QPSK modulator takes 1 bit from each stream to construct the phase constellation having four phases, also known as "symbols," where each symbol represents 2 bits. The symbol rate is therefore reduced by a factor of 2. The QPSK modulator has four phases or four symbols, 2 bits/symbol as shown in the figure.

Therefore, the QPSK modulator has the following specifications:

• Four phases or four symbols
• 2 bits/symbol

The above specifications govern the transmission bandwidth, as we shall see later.

11.4   8PSK Modulation

In 8PSK, the input raw data, having a bit rate R_{b1}, is encoded by a rate r $(r < 1)$ encoder. The encoded data, having a bit rate $R_{b2} = R_{b1}/r$, is serial to parallel converted into three parallel streams. The encoded bit rate, now reduced in speed by a factor of 3, is modulated by the 8PSK modulator as shown in Fig. 11.6.

The 8PSK modulator takes 1 bit from each stream to construct the phase constellation having eight phases, also known as "symbols," where each symbol represents 3 bits. The symbol rate is therefore reduced by a factor of 3. The 8PSK modulator has eight phases or eight symbols, 3 bits/symbol as shown in the figure.

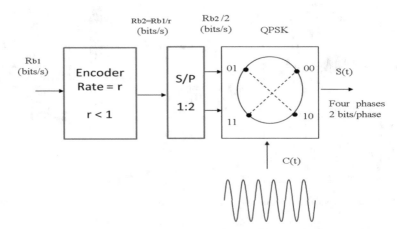

Fig. 11.5   QPSK signal constellation having four symbols, 2 bits per symbol

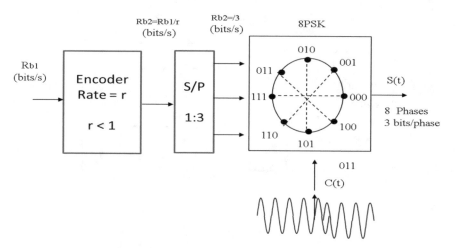

Fig. 11.6 8PSK signal constellation having eight symbols, 3 bits per symbol

Therefore, the 8PSK modulator has the following specifications:

- Eight phases or eight symbols
- 3 bits/symbol

The above specifications govern the transmission bandwidth, as we shall see later.

11.5   16PSK Modulation

In 16PSK, the input raw data, having a bit rate R_{b1}, is encoded by means of a rate r ($r < 1$) encoder. The encoded data, having a bit rate $R_{b2} = R_{b1}/r$, is serial to parallel converted into four parallel streams. The encoded bit rate, now reduced in speed by a factor of four, is modulated by the 16PSK modulator as shown in Fig. 11.7.

The 16PSK modulator takes 1 bit from each stream to construct the phase constellation having 16 phases, also known as "symbols," where each symbol represents 4 bits. The symbol rate is therefore is reduced by a factor of 4. Therefore, the 16PSK modulator has 16 phases or 16 symbols, 4 bits/symbol as shown in the figure.

Therefore, the 16PSK modulator has the following specifications:

- 16 phases or 16 symbols
- 4 bits/symbol

The above specifications govern the transmission bandwidth, as we shall see later.

Table 11.1 shows the number of phases and the corresponding bits per phase for MPSK modulation schemes for $M = 2, 4, 8, 16, 32, 64$, etc.

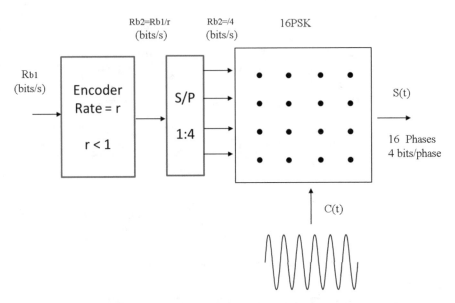

Fig. 11.7 16PSK signal constellation having 16 symbols, 4 bits per symbol. Here, each symbol is represented by a dot, where each dot represents 4 bits

Table 11.1 MPSK modulation parameters. $M = 2, 4, 8, 16$, and 32

Modulation	Number of phases φ	Number of bits per phase
BPSK	2	1
QPSK	4	2
8PSK	8	3
16	16	4
32	32	5
64	64	6
:	:	:

11.6   PSK Spectrum and Bandwidth

In wireless communications, the scarcity of RF spectrum is well-known. For this reason, we have to be vigilant about using transmission bandwidth in error control coding and modulation. The transmission bandwidth depends on:

- Spectral response of the encoded data
- Spectral response of the carrier phase
- Modulation type

   Let's take a closer look.

11.6.1 *Spectral Response of the Encoded Data*

In digital communications, data is generally referred to as a nonperiodic digital signal. It has two values:

- Binary-1 = high, period = T
- Binary-0 = low, period = T

Also, data can be represented in two ways:

- Time domain representation
- Frequency domain representation

The time domain representation (Figure 11.8a), known as non-return-to-zero (NRZ), is given by

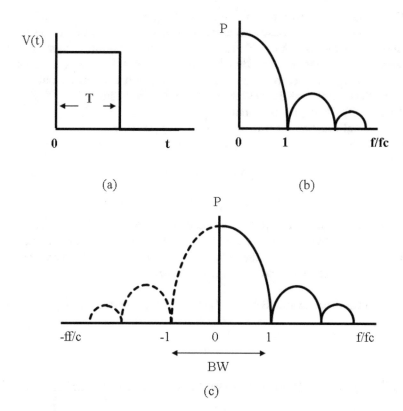

Fig. 11.8 (**a**) Discrete time digital signal. (**b**) Its one-sided power spectral density and (**c**) two-sided power spectral density. The bandwidth associated with the non-return-to-zero (NRZ) data is $2R_b$, where R_b is the bit rate

$$V(t) = V \quad < 0 < t < T$$
$$= 0 \quad \text{elsewhere} \tag{11.5}$$

The frequency domain representation is given by "Fourier transform":

$$V(\omega) = \int_0^T V.e^{-j\omega t}\,dt$$

$$|V(\omega)| = 2\left(\frac{V}{\omega}\right)\text{Sin}\left(\frac{\omega T}{2}\right) = VT\left[\frac{\text{Sin}(\omega T/2)}{\omega T/2}\right] \tag{11.6}$$

$$P(\omega) = \left(\frac{1}{T}\right)|V(\omega)|^2 = V^2 T\left[\frac{\text{Sin}(\omega T/2)}{\omega T/2}\right]^2 \tag{11.7}$$

Here, $P(\omega)$ is the power spectral density. This is plotted in Figure 11.8b. The main lobe corresponds to the fundamental frequency, while side lobes correspond to harmonic components. The bandwidth of the power spectrum is proportional to the frequency. In practice, the side lobes are filtered out, since they are relatively insignificant with respect to the main lobe. Therefore, the one-sided bandwidth is given by the ratio $f/f_b = 1$. In other words, the one-sided bandwidth $= f = f_b$, where $f_b = R_b = 1/T$, T being the bit duration.

The general equation for two-sided response is given by

$$V(\omega) = \int_{-\infty}^{\infty} V(t).e^{-j\omega t}\,dt$$

In this case, $V(\omega)$ is called the two-sided spectrum of $V(t)$. This is due to both positive and negative frequencies used in the integral. The function can be either a voltage or a current. Figure 11.8c shows the two-sided response, where the bandwidth is determined by the main lobe as shown below:

$$\text{Two sided bandwidth (BW)} = 2R_b (R_b = \text{Bit rate before coding}) \tag{11.8}$$

11.6.2   Spectral Response of the Carrier Before Modulation

A carrier frequency is a sinusoidal waveform, which is periodic and continuous with respect to time. It has one phase component. For example, the sine wave is described by the following time domain equation:

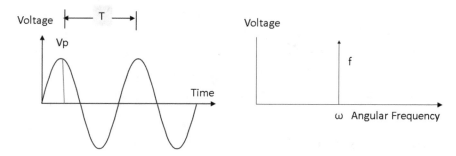

Fig. 11.9 High-frequency carrier response

$$V(t) = V_p \text{Sin}(\omega_c t) \tag{11.9}$$

where

Vp = peak voltage

- $\omega_c = 2\pi f_c$
- f_c = carrier phase in Hz

Figure 11.9 shows the characteristics of a sine wave and its spectral response. Since the phase is constant, its spectral response is located in the horizontal axis, and the peak voltage is shown in the vertical axis. The corresponding bandwidth is zero.

11.6.3   BPSK Spectrum

In BPSK, the input is a digital signal and it contains an infinite number of harmonically related sinusoidal waveforms. This is given by (see Sect. 6.4.1)

$$|V(\omega)| = 2\left(\frac{V}{\omega}\right)\text{Sin}\left(\frac{\omega T}{2}\right) = VT\left[\frac{\text{Sin}(\omega T/2)}{\omega T/2}\right] \tag{11.10}$$

$$\textit{Equation} \tag{11.11}$$

Here, $V(\omega)$ is the frequency domain representation of the input digital signal, which has a $\text{Sin}(x)/x$ response that governs the phase of the carrier frequency.

With $V(t) = m(t)$, we write

$$S(t) = A_c \text{Cos}[\omega_c t + \beta\, m(t)] \tag{11.12}$$

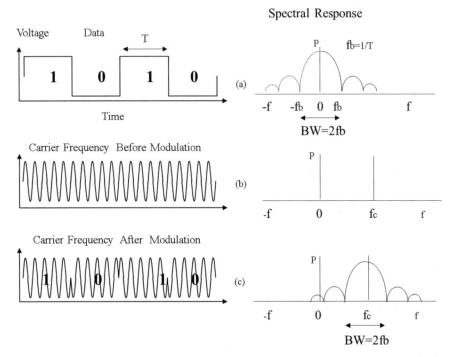

Fig. 11.10 BPSK bandwidth. (**a**) Spectral response of NRZ data before modulation. (**b**) Spectral response of the carrier before modulation. (**c**) Spectral response of the carrier after modulation

where β is the phase deviation index of the carrier and $m(t)$ has a $Sin(x)/x$ response, which is given by

$$m(t) = VT \left[\frac{Sin(\omega mT/2)}{\omega mT/2} \right] \qquad (11.13)$$

Therefore, the spectral response after BPSK modulation also has a $Sin(x)/x$ response, which is the shifted version of the NRZ data, centered on the carrier frequency f_c, as shown in Fig. 11.10. The transmission bandwidth associated with the main lobe is given by

$$\text{BW (BPSK)} \approx 2R_{b2}/\text{Bit per Phase}$$
$$\approx 2R_{b2}/1 \approx 2R_{b2} \qquad (11.14)$$

where R_{b2} is the coded bit rate (bit frequency). Notice that the BPSK bandwidth is the same as the ASK bandwidth. It may be noted that the higher-order spectral components are filtered out.

Problem 11.1

Given:

- Uncoded input bit rate: $R_{b1} = 10$ kb/s
- Code rate: $r = 8/15$
- Carrier frequency $f_c = 1$ MHz
- Modulation: BPSK, QPSK, 8PSK, and 16PSK

Find:

(a) Coded bit rate R_{b2}
(b) BPSK bandwidth (BW)
(c) QPSK bandwidth (BW)
(d) 8PSK bandwidth (BW)
(e) 16PSK bandwidth (BW)

Solution:

(a) Coded bit rate: $R_{B2} = R_{B1}/r = 10$ kb$/(15/8) = 18.75$ kb/s
(b) BPSK bandwidth: BW $= 2R_{B2}/1 = 2 \times 18.75 = 37.5$ kHz
(c) QPSK bandwidth: BW $= 2R_{B2}/2 = 2 \times 18.75 /2 = 18.75$ kHz
(d) 8PSK bandwidth: BW $= 2R_{B2}/3 = 2 \times 18.75/3 = 12.5$ kHz
(e) 16PSK bandwidth: BW $= 2R_{B2}/4 = 2 \times 18.755/4 = 9.375$ kHz

Note: Higher-order PSK modulation is bandwidth efficient.

Problem 11.2

Given:

- Input bit rate $R_{b1} = 10$ kb/s
- Code rate $r = 1/2$
- Modulation: BPSK, QPSK, 8PSK, and 16PSK

Find:

(a) Bit rate after coding R_{b2}
(b) Transmission bandwidth BW

Solution:

(a) $R_{b2} = R_{b1}/r = 10$ kb/s $\times 2 = 20$ kb/s

(b) Transmission bandwidth:

- BPSK BW $= 2R_{b2}/$bits per symbol $= 2 \times 20$kb/s/1 $= 40$ kHz
- QPSK BW $= 2R_{b2}/$bits per symbol $= 2 \times 20$kb/s/2 $= 20$ kHz
- 8PSK BW $= 2R_{b2}/$bits per symbol $= 2 \times 20$kb/s/3 $= 13.33$ kHz
- 16PSK BW $= 2R_{b2}/$bits per symbol $= 2 \times 20$kb/s/4 $= 10$ kHz

Note: Higher-order PSK modulation is bandwidth efficient.

11.7   Conclusions

- This chapter presents PSK modulation and its attributes.
- Numerous illustrations are provided to show how the phase of the carrier changes in discrete levels in accordance with the input digital signal, while the amplitude of the carrier remains the same.
- The Fourier transform is used to derive the spectral components and PSK bandwidth is calculated.
- These materials have been augmented by diagrams and associated waveforms to make them easier for readers to grasp.

References

1. D.R. Smith, *Digital Transmission Systems* (Van Nostrand Reinhold Co, New York, 1986). ISBN: 0442009178
2. W. Leon, I.I. Couch, *Digital and Analog Communication Systems*, 7th edn. (Prentice-Hall, Inc., Englewood Cliffs, 2001). ISBN: 0-13-142492-0
3. B. Sklar, *Digital Communications Fundamentals and Applications* (Prentice Hall, Englewood Cliffs, 1988)
4. S. Faruque, *Radio Frequency Channel Coding Made Easy* (Springer, Cham, 2016). ISBN: 978-3-319-21169-5
5. C.C. George et al., *Error Correction Coding for Digital Communications* (Plenum Press, New York, 1981)
6. G. Ungerboeck, Channel coding with multilevel/multiphase signals. IEEE Trans. Inf. Theory **IT28**, 55–67 (1982)
7. S. Lin, D.J. Costello Jr., *Error Control Coding: Fundamentals and Applications* (Prentice-Hall, Inc., Englewood Cliffs, 1983)
8. R.E. Blahut, *Theory and Practice of Error Control Codes* (Addison-Wesley, Reading, 1983)

9. J.H. van Lint, *Introduction to Coding Theory*, GTM 86, 2nd edn. (Springer-Verlag, New York, 1992), p. 31. ISBN 3-540-54894-7

10. F.J. Mac Williams, N.J.A. Sloane, *The Theory of Error-Correcting Codes* (Amsterdam, North-Holland, 1977), p. 36. ISBN 0-444-85193-3

11. W. Huffman, V. Pless, *Fundamentals of Error-Correcting Codes* (Cambridge University Press, Cambridge, 2003). ISBN 978-0-521-78280-7

12. P.M. Ebert, S.Y. Tong, Convolutional Reed-Soloman codes. Bell Syst. Tech., 729–742 (1968)

13. R.G. Gallager, *Information Theory and Reliable Communications* (Wiley, New York, 1968)

14. A. Kohlenbero, G.D. Forney Jr., Convolutional coding for channels with memory. IEEE Trans. Inf. Theory **IT-14**, 618–626 (1968)

15. M.A. Reddy, S. M, Further results on convolutional codes derived from block codes. Inf. Control. **13**, 357–362 (1968)

16. S.M. Reddy, *A Class of Linear Convolutional Codes for Compound Channels*, Technical Report (Bell Telephone Laboratories, Holmdel, 1968)

Index

© The Editor(s) (if applicable) and The Author(s), under exclusive license to
Springer Nature Switzerland AG 2021
S. Faruque, *Free Space Laser Communication with Ambient Light Compensation*,
https://doi.org/10.1007/978-3-030-57484-0

Printed in the United States
by Baker & Taylor Publisher Services